Kh

By Richard Allen Khalar

For information regarding permission email: Richard Khalar at
R.KHAL42@GMAIL.COM

Edition: 3

ISBN-13: 978-1727433906
ISBN-10: 1727433904

Printed in the USA by Pro Print in Superior, WI

Table of Contents

Prologue

Why I write?

Dad fought across Europe during World War Two (WWII) in a halftrack armed with quad .50 caliber Anti-Aircraft machine guns. Later in his life he told me a few stories about his war experiences. (Appendix 1) I wish I had pressed him for more. Many of my uncles also fought in WWII, others served after the war but I know little or nothing about their service. One uncle landed in the second wave on Omaha Beach on D-day. I once asked him about it. He refused to talk about it. So much family history lost.

I write because I want to pass my military and life experiences on to my family and friends.

I'm not much of an author. My attempt here is simply to capture my military and life experiences on paper. You will probably find spelling, word structure, and grammar errors throughout the stories. Also, you may read acronyms and military terms you may not understand. Read, but be gentle, and when you find an error, please just smile. If you want to pass a suggestion or comment to me on what you read, please do.

I retired in 1999 after serving 30 years in the Air Force. Many years later, after I retired, my college Air Force Reserve Officer Training Corps class organized a reunion. As part of the ROTC reunion spin up, we were all asked for an "Unofficial Bio" (Appendix 2) about our experiences. They didn't want the "Official Air Force Biography" (Appendix 3) which only has dates, job titles, ranks, and award lists. They wanted a couple of pages about our AF "experiences."

At that point in my life, I had not given much thought about my Air Force life. I had retired years earlier and just walked away. I didn't sense anything special about my experiences as they were just part of my normal day-to-day routine life.

But then, I sat down and to my surprise, quickly filled up two pages of my Unofficial Bio experiences, and then stopped. There were many more events that I could have added. Memories flooded back. I leaned back and re-read the two pages. Wow, I thought, did I really live this life? Some of it was surreal. Was that really me? And I suddenly realized, I couldn't believe that I was still alive!

After the reunion, my two page "Unofficial Bio" sat around for several years. Then my cousin Ken shared with me several short stories that he wrote. His works triggered me to write my first experience account, "Close Encounters," a story about a B52D model combat sortie over Laos.

Since I retired for good in April 2015, over three years ago, my writing has accelerated. I write what I recall. It's my best attempt and my intention to tell these events accurately. So how do I know? Well, I have my Air Force records. I have every Order cut sending me places, each Travel Voucher I filled out documenting my travel, and all my Officer Effectiveness Reports written by my superiors about my accomplishments. I also have my yearly calendars with events, dates, and times. Finally, I have my flight records and logbooks that document my flights listed. So, I know where I've been, when, and why.

I don't think I'm anyone special. Having said that, these are my experiences. Many of the events I've written about were stressful situations. I read through them and wonder now how I survived the 30 years. I'd like to think I'm just good, but in reality, either I'm really good, lucky, or someone was watching over me. Probably all three.

Please understand that the stressful B52 situations I write about are representative of all B52 crewmembers. You are reading about the lives of those that fly and flew the B52.

Stressful situations aside, trust me, I've experienced boredom, hours and hours, of shear boredom. How boring would it be to write about boring experiences?

I've made my share of mistakes. You will read about many of them. Obviously, I survived them.

Also, there was, and still is, lots of comradery in my world. As in, "Band of Brothers" comradery. Social media allows me to maintain contact with thousands of B52 associated folks, including dozens of crewmembers I've served with.

There were also lots, and I mean lots, of humor in my military life. You might notice that I liked to screw with people, especially the Friggen New Guy (FNG).

These pages reflect the kinds of happening that occurred to my many fellow military members on a daily basis, just ask them. Maybe you have a veteran friend or family member that needs to be encouraged to write about their experiences. Encourage them.

I write events in chronological order from grade school through my final retirement. I don't write much about my family. I have great love and admiration for my family. My family is a very personal issue to me. Obviously, they know me through the course of normal family interactions. There is much of my life that is not known to them. The information in this work will be new to them. My family and friends will know me better.

I wish Dad, and my Uncles, had passed their World War II and military experiences on to the family. Here is a picture of my dad (Joe), uncles, aunts, and grandparents.

Larry, Gram, Steve, Tom, Marty, Floyd, Joe (Dad), Clarence, Art, Alice, Fred, Eugene, Albert, Gramps.

Chapter 1 - The Early Years

First Flight

My first aircraft flight occurred when I was in the seventh grade. There was a small airshow at the Town of Highland airport. A pilot was selling rides in a Cessna 172 (single engine prop with 4 seats) for a penny a pound. I got on the scale and bought a ride for $1.50. We took off but didn't go far. From my first bird's eye view of the world, from above I looked north across the Brule River swamp. There was highway S, and Sunfish Lake, and there was our house! Further north was Lake Nebagamon. Then 30 miles farther north, I see Lake Superior! Wow, this was what birds see! This was flight!

Town of Highland is located in the northern tip of Wisconsin about 40 miles southeast of the City of Superior. It isn't really a town, rather, it's just a land mass. It was named Highland because it was the high land to the south of Lake Superior. It is forest covered with many lakes scattered throughout. We were country people. Back then, the population was about 85, now it's about 250. The Highland airport's two runways were grass. There were no buildings or structures around it.

8th Grade Class. Two members missing

After the Highland one room schoolhouse closed, all the kids were bussed five miles away to the grade school in Lake Nebagamon. The grade school in Lake Nebagamon was small. There were four rooms with two grades in each room. Eleven students were in my eight-grade class. We were organized in two rows. The two rows next to us were the 7th grade.

After grade school, I attended Northwestern High School in Maple, Wisconsin, graduating in 1965. It was a small school. There were 80 in my class with about 380 in the school. Kids bussed in from eight surrounding towns. The school paper was called the Octagon. I spent about two hours a day on a bus.

Highland Humor

I may have learned my sense of humor from my dad. Ask any of my cousins, Dad had them all convinced there were vampires living in the basement. They didn't want to go down into the basement.

A couple of years ago our two grandsons came to our summer home in Wisconsin, the house I grew up in. Rumors of trolls living in the basement caused quite a stir. We made a "capture the troll" plan. They carefully descended the stairs down into the basement. They were armed with flashlights and a large fishing net. The plan was to locate and net the troll. Unfortunately, no trolls were found after the careful search.

We didn't have Snipe in the area so we didn't snipe hunt. There were, however, at least three creatures in the Highland forests to be avoided: Snow Snakes, Side Hill Gouges, and the ever-nasty Swamp Slogger.

City slickers from the Milwaukee area would occasionally come up and join our crew to deer hunt. They somehow learned about the Snow Snakes. Snow Snakes were white so you couldn't see them in the snow. If you walked with your arms hanging down, the snakes could strike up and attach onto one of your fingers. They were extremely poisonous so you had to act fast. You had to quickly put your hand up against a tree trunk and with your hunting knife, quickly hack off the finger with the snake attached. We noted all the city slickers walked through the snow with their rifles and hands chest high. There were no hanging arms.

The Side Hill Gouge lived on hill sides. They only moved in the same direction around a hill top. Over time their two uphill legs became shorter than their two downhill legs. So, if you were chased by one, run up or down hill and not around the hill. The reason is because of the legs they couldn't run in a straight line and catch you.

The Swamp Slogger was the most dangerous creature in Highland woods. During my post freshman college summer, my friend, Dave, and I ran a pulp peeling operation. He and I cut down poplar trees. A number of friends and relatives peeled the bark off the trees. Rumor has it that one of my cousins was, in fact, chased through a swamp by a Swamp Slogger!

Work Experience

I worked a number of jobs while attending high school and college. My first job was peeling pulp in the forest. Poplar trees were cut down in the spring when sap began to run. We then manually peeled the bark off the tree. The tree (pulp) was eventually hauled off to the paper mill in Superior, WI, to be made into paper products.

A crew consisted of a sawyer (cuts the trees down with a chainsaw), and whoever showed up to peel (usually 4 or 6). I worked using three tools: An axe, an 8-foot measuring stick, and a peeling spud (1½ foot length of old car spring). The sawyer dropped a tree. Ideally, it didn't lay directly on the ground because it was harder to peel the bottom bark off. Starting at the butt, moving toward the tree top using the marking stick and axe, I chopped a mark into the tree every 8 feet. I also used the axe to lop off branches that were near the top. There were usually 5 sticks in a tree. Standing at the tree top, the axe and marking stick were thrown back toward the tree butt. The spud was stuck through the bark and was pushed around between the bark and the tree thus peeling the bark off.

At days end, what remained was a maze of white overlapping peeled trees. The trees laid there through the summer drying out. In the fall, a sawyer would work through the maze cutting up the 8-foot marked lengths. The logs were then pulled into small piles and a caterpillar pulling a skidder would come through the woods, load the logs, and bring them out of the woods to be loaded onto trucks.

We were paid 7 cents for each stick peeled. A stick was an 8-foot length of the tree log. On a good day I could make $15-$20. My best day ever was 300 sticks or $21.

Leroy was a teacher at Northwestern High School. During the summer he worked in the woods peeling pulp with us for his brother Lyle. He and I had a daily competition who could peel the most pulp. I only beat Leroy one day! Many years later I had the chance to meet him again. He was elderly, ill, and could barely walk or talk, but he proudly reminded me that I only beat him once!

Glen was a big strong Norwegian. He was a partner with Lyle. He called me "the tennis shoe kid" because I wore tennis shoes in the woods. When my tennis shoe sides ripped out I wrapped duct tape around them. One time the crew was eating lunch and we started to wrestle. Actually, I just jumped onto his back, he just reached over his shoulder, grabbed me, and threw me to the ground.

I learned how to work in the woods. There were no hourly wages. You were paid by the number of sticks you peeled. In this world, there were no unions, no paperwork, no health insurance, just a hand shake, and show up. Honesty was a cornerstone. I would show up at Lyle's door, tell him how many sticks I peeled that week, and he wrote me a check.

Thinking back, my first attempt peeling pulp was years earlier with my uncle Larry. He had just returned from Germany. He had been in the army. One day he says, "Come on." He leads me down Francis Willard Road to the field. We are carrying a bowsaw and a couple of large screwdrivers. Along the field edge he cut down a poplar tree with the bowsaw. We then tried to peel the bark off it with the screwdrivers. Didn't happen. I had no idea of what we were doing. He apparently knew something about it but the timing wasn't right. Bark is peeled during the spring when sap is running.

The manual peeling of pulp ended many years ago. Now a big tracked vehicle pulls up to a tree, grabs it, then cuts it off near the ground. The tree is then lowered down to be parallel the ground and the entire tree is then fed through a wood chipper. The chips were transported to the paper factory. I wonder how kids learn to work today?

Like a bat hitting a rag doll

Working in the woods was a hazardous occupation. A lapse of attention could kill you. That is, unless you're just lucky.

I had an uncle who, during World War II, landed on D-day in the second wave onto Omaha Beach, Normandy. He survived, lucky? Years later he worked in the woods cutting down trees. I heard that he once broke a leg and dragged himself through the woods to a road. Laying on the road side he flagged down a car and help. Lucky again?

I started to work in the woods when I was in high school by peeling the bark off of poplar trees. Later on, I went from peeling pulp to using a chain saw to cut trees down. I've cut thousands of trees down. Cutting poplar trees without peeling was called cutting "rough." It was usually done in the fall and winter when the sap wasn't flowing so they couldn't be peeled. Cutting pulp logs without being peeled was worth about $23 a cord.

During my junior year in college, my uncle offered me the pulp on his property. I started cutting it. One day I was in the woods alone cutting poplar trees down on the west side of Sunfish Lake. I made the final cut on a tree and as it started to fall I looked down range again. I saw something I missed on my precut look! I immediately knew I was in big trouble. About midway down range there were two trees each tilted past each other forming an X. The closest tree leaned right, the farther tree leaned to the left. I knew when the falling tree hit the X the tree butt was coming my way! It all happened so fast. I just had time to throw my running chainsaw into the air as the butt bounced straight up about shoulder high then slightly away from me parallel to the ground, then it slapped me in the stomach like a baseball bat hitting a rag doll.

It hit me at the stomach level with my head/arms over it and my legs wrapped under it. Luckily, it slapped and threw me about 5 feet from where it dead dropped to the ground. Had it dropped when I was folded over and under it I would still be there. Most likely the tree would have dropped on me and crushed my pelvis and/or legs and pinned me to the ground. If I hadn't seen the danger and thrown the chainsaw, the running saw would have been slammed into me! Who knows what would have resulted in that. I was in the woods alone and my folks didn't know where I was. I laid there in a daze for a few minutes and to figure out my injuries. I couldn't believe there were no injuries. I called it a day.

That is the story of my life. I should have died that day, but I was just lucky. I have lived many lucky days in my life, luck is the story of my life.

College Years

I attended college at the University of Wisconsin, Superior, 1965-1969. I believe I was the first Khalar to attend and graduate from college. My dad completed 6th grade. Mom graduated high school. They were both wonderful, caring, intelligent people.

AFROTC

We all make decisions on a daily basis. Taking either a Physical Education class or Air Force Reserve Officer Training Corps (AFROTC) was mandatory in the freshman year. I joined ROTC. The Vietnam War was in progress. Being drafted into the military was a real thing. You were either in college or you were drafted. Four years was a hard number, graduate in four years or you were drafted. If a man was short of one credit in any school year he was drafted. I graduated in four years. Besides, I figured after completing college I'd be drafted. I joined ROTC so I would be a commissioned officer in the Air Force. If drafted I had no way to know what service I would be assigned.

My freshman college year I lived on campus in Ostrander Hall on the first floor. During my sophomore year I lived at home and commuted the 30 miles to/from college. Over the sophomore summer I ran into Roger at a Saturday night dance in Lake Nebagamon. Roger and I met during our freshman year living in Ostrander Hall. He asked me if I wanted to share an apartment beginning our junior year. I said, "Yes!" During my junior and senior years, we lived in the apartment on 18th street, about four blocks from the college.

It was a three-story house. We lived on the top floor. The apartment was small with a bathroom, kitchen, two bedrooms and a living room. The building was owned by the Cronk's. Dr. Cronk lived on the second floor with his wife and mother (Granma Cronk). He was a professor at the School. Another university professor lived on the first floor. Mrs. Cronk was a teacher at the Port Wing High.

Our rent was $50 a month plus half the cost of the heat for the entire building. My split cost during the summer was about $17 a month. My winter cost with half the building heat factored in was $50 a month! During the junior year my roommates were Roger and Ralph. Ralph switched colleges during the junior summer transferring to Stevens Point College for his senior year. Roger and I thought just the two of us would be in the apartment for our senior year.

I'm pointing to my future 2nd Lieutenant Insignia

Then on the first day of school Arron knocked on our door. His prior year roommates wouldn't let him move back with them for the senior year? He needed a place to live and asked if he could stay with us. We let him move in with us. (We should have wondered why his earlier roommates didn't let him back in with them.)

Someone told me that as a student I could apply at the county courthouse for food stamps. I applied. I then took my book of food stamps to a grocery store. When I got to the checkout, I handed out the food stamps I felt so embarrassed that I never used them again. I decided that I could get by without them.

While in college, I worked on the Duluth/Superior harbor docks as a Stevedore, Longshoreman, and Night Watchman. The pay was $2.40 an hour. As a Stevedore, I unloaded railroad box cars filled with 1,400 one hundred-pound bags of dried milk. Railroad box cars were parked next to doors along the length of the warehouse. Each of us would unload one boxcar a day.

The first step in the days work was to break the metal seal on the railcar door and open the door. The 100 lb. bags were stacked to about shoulder height in the car. I pulled and carried one bag out at a time and placed it on a wooden pallet. I stacked the bags on the pallet up to about shoulder high. Then I drove a forklift to haul the loaded pallet into a storage location in the warehouse. The pallets remained there stacked up about three pallets high waiting for a ship to load.

One day after I first began working there, I unloaded my boxcar too fast. The boss opened a second boxcar for me. Boy, were the union guys mad at me! They had to help me unload the second car. I learned to pace myself (slow down) to the union speed.

Once in awhile I went into the ship's cargo hold as a longshoreman. The loaded pallets were moved from the warehouse and lowered into the ship. We carried the 100 lb. bags from the pallets and spread them throughout the hold.

One day the guy next to me says, "Sit down." I looked around to see the entire crew sitting down. We sat there for a minute and I asked, "Why are we sitting?" He pointed to another man and said, "The union steward said to sit down." I then asked the guy, "Why?" He pointed to a light bulb that was burned out. It was one bulb on a long string hung across the hold. The eight of us sat there for about 30-40 minutes until the light bulb was replaced. That one bulb out made no difference in the lighting in the hold. Eight guys, 30-40 minutes, no work, welcome to the world of the unions.

As night watchman, I stayed in the dock warehouse overnight when a ship was docked there. I had to watch the place. The ship's crew came and went via taxi all night. Occasionally a "lady of the night" would arrive and board the ship.

One night a ship arrived and I was standing on the dock as a cargo ship slowly slid along the dock next to me. A crewmember leaned over the railing above and calls down, "Where do you want me to tie up?" What do I know, so I say, "Right there, that's perfect?" Luck was with me because it turned out that it was the right place.

In 1968 I was a junior in college. The ROTC

Professor of Aerospace Studies was a Lt Col. Being an Air Force pilot, he was required to fly at least 4 hours a month to continue to remain flight current and to receive flight pay. He drove up to the Duluth Air Base to fly a T33 (T-Bird).

A T-Bird is a two-seat single engine jet. He would take three cadets with him and give each cadet a flight. My first jet flight! We took off and were soon half way down the state of Wisconsin. I recall thinking that it took me 45 minutes to drive from the Superior campus to the Duluth Air Base and we had already flown half way down the state in less time. Years later that same ramp is where Cirrus Aircraft setup business. I have flown onto that ramp many times since.

First Jet flight, a T33

A car load of five cadets traveled from Superior to the Duluth Air Base hospital (at Duluth airport) for commissioning/flight physicals. It was winter, a sunny bright day, and the ground was covered with snow. The last item checked during our physicals was our vision. Our eyes were still dilated when we five completed the physicals and were released. We opened the hospital door to walk to the parking lot. With dilated eyes it was way too bright to see. We walked in line with a hand on the persons shoulder in front of us to the car. I do not remember who else was there or who drove.

At some point during the physical I was asked to sit on a table with my back against the wall. My sitting height was measured. I was then told that my sitting height was too tall to be an Air Force pilot and attend pilot flight school.

The next day I told the Professor of Aerospace Science that since my sitting height was too tall to go to pilot training, I would finish the current semester in ROTC then not return. A month or so later the he told me that the sitting height had been increased and that I now met the pilot sitting height criteria. (More on that later)

Field Training

All ROTC cadets were required to attend and complete a 4-week field training camp between the junior/senior years. (Later it was changed to between the sophomore and junior years.) I was assigned to the camp at Gunter AFB, Montgomery, AL. There were 6 Superior University cadets assigned to the Gunter camp. We would spend August, 1968, at the camp.

We drove south in Ed's new Monte Carlo from Superior, Wisconsin, to Gunter AFB. There were 6 of us in the car, Ed, Bill, Jim, Steve, Dave, and me. We took turns driving. On the way, we stopped at a country club in the Chicago area to visit my younger sister Bev. She worked there. We then stopped at Fort Campbell, Kentucky, so Steve could visit a friend (Tim) and deliver brownies from Tim's mom to him. I didn't know him at the time but later Tim married my sister, Bev!

At the ROTC training camp at Gunter AFB, there were 25 cadets in each flight. There were 10 flights so 250 cadets in the camp. Training was an intense 4 weeks. Each flight had an active duty Tactical (TAC) Officer to oversee it. My TAC officer was Major Strong, a KC135 tanker navigator. (Years later I ran into him at a flight line snack bar on Utapoa, Thailand, Air Base. It really is a small world.)

I reported to him when I arrived and he assigned me as the Flight Athletic officer. I don't know why he selected me for the position? I've wondered was the assignment just out of the blue, or did he have information about cadets? I was, in fact, very physically fit. I held the position for the entire month.

Field Training was an opportunity for cadets to demonstrate how to follow and how to lead. Flight leadership positions were rotated throughout the 4 weeks. One of my fellow flight cadets was assigned as the flight

Standardization Officer. He briefed the flight on what we needed to do to comply. I learned about "Standardization" that first night. Everything had a correct place to be. Among many things, believe it or not, my underwear was to be rolled/folded/stacked neatly so the folds resulted in a visible smile! We tried to arrange our rooms according to the published standardization plan.

But the standardization plan didn't work in the rooms on both sides of the hall. This was a problem so the flight met. Somehow, I was selected to go and tell the TAC officer (Major Strong) that his/their published standardization plan didn't work! I don't recall why the cadet assigned as Standardization Officer or the Flight Commander didn't do it?

Anyway, I reported to Major Strong and told him it was impossible to comply with the standardized guidance. He said, "wait here," and left the room. He came back after awhile and said, "Your right, now go back to your flight." Question. Is it ok to tell your boss on the first day that he is wrong!

Later, that first evening we had a fire drill. The fire alarms went off. We fell out and 250 cadets formed up in front of the barracks. It was an exercise. It was raining and we were all soaked by the time we were dismissed back to barracks. We then cleaned rooms and buffed the hall floor.

So, I'm the Flight Athletic officer. The next morning, it was still dark, we assembled by flight and the TAC officer asked me come forward and to lead the flight in the 5BX. I'm standing in front of 24 other cadets on morning one. I had no idea of what the "5BX" was! Never heard of it. The TAC officer asked if anyone knew what it was. One of my fellow flight members came out and stood next to me and led the exercise sequence. How embarrassing was that? The 5BX was a Canadian physical fitness exercise sequence (5 exercises) designed for their pilots. Where they got the name 5BX was never made know to me, but the US Air Force adopted it.

There were three cadets in each room. I hated the pop sound that the loud speakers made when turned on just before revelry played each morning. Up we jumped and put on our PT uniforms. We fell into formation and marched in the dark to the athletic field where we did the 5BX exercise in mass. Then a run by flight was made. Flight Athletic officers took turns leading the 250 cadet camp exercises from a raised platform. I took my turn on the raised platform and led the camp. I learned the 5BX exercise sequence quickly!

We always looked for ways to save time. We each had a padlocked lockbox for non-standardization and personal items. To open it took four lock number changes. Before going to bed I made three number changes so in the morning only one would open the lock. I learned if I slept on top of the bunk bed blankets, I saved bed making time in the morning. The flight only used two of the six urinals and two of the six toilets. That was eight fixtures that we didn't have to clean to prepare for the morning inspection.

I had a T33, T-Bird, flight out of Maxwell AFB. My pilot was a navy officer attending one of the schools at Maxwell, Air Force Base. We shot and qualified on the .38 Smith & Wesson Combat Master Piece. (A few years later I carried a Smith & Wesson in the real world on 78 combat sorties in South East Asia.)

We did a survival night in the woods on Gunter, AFB. We each had two blankets. I made the mistake of sweeping all the leaves and pine needles off the ground. I laid on the bare ground. I woke up before sunup freezing! I wrapped the blankets around me and sat up shivering until morning. Years later I attended AF Survival School in Spokane, WA. Leaves and needles would have been insulation under me!

Several days later, there was a Field Day Competition where all flights competed against each other in all sorts of athletic events. As Flight Athletic officer I selected who would compete in each activity. There was an obstacle course race. Each flight competed with a five-man team. Four cadets would carry a cadet on a stretcher across an obstacle course. The best time won. We won! I picked the Tug of War members and a cadet to yell us to win. We won! There was an event that timed how long it took for the flight to swim across the swimming pool. Ok, not everyone is a great swimmer. I briefed the flight that the first across to reach back and make a chain to reach back to the slower swimmers and pull them across. We won. We did really well!

At the end of the 4-week camp a Dining-In was held. That's a formal military dinner. I was naïve and didn't know

The Vice Commandant Award

anything about the ROTC award system. The Camp Commandant (an AF Colonel) began to recognize cadets with individual awards. The "Vice Commandant Award" was given to the top cadet in each flight. I couldn't believe it when he announced my name! Totally blew me away! I learned from this unexpected experience that I was competitive.

Later when we were alone for a few minutes the Camp Commandant asked me why I didn't compete in

VICE COMMANDANT'S AWARD

OUTSTANDING ACHIEVEMENT

Be it known that Air Force ROTC Cadet

RICHARD A. KHALAR

having demonstrated distinguished achievement in all aspects of

Air Force ROTC Field Training

and having been selected as the most outstanding individual in

Flight F, GUNTER Air Force Base,

is hereby designated an OUTSTANDING CADET.

E.C. Murphy, Col USAF
Field Training Commander

Date 5 September 1968

Colonel, USAF
Vice Commandant, AFROTC

Vice Commandant Award Certificate

certain athletic activity. I told him that, although that event should have been mine, one of my flight mates had asked me if he could compete in the activity. I said ok. He didn't do well. My take is that if I had competed in the event, I might have had a shot at being selected as number one in the camp. Who knows?

(27 years later I was the AFROTC Colonel Commandant for a 4-week Field Training Camp that held 520 cadets. It was at Lackland AFB, San Antonio. At the time it was the biggest AFROTC camp ever!)

Our new Professor of Aerospace Studies had just returned from an assignment in Okinawa and had a VW Van shipped back to the States. It needed to be driven up to Superior. Three of us drove it north. We stopped in Milwaukee to visit my uncle. Uncle Larry worked at the Pabst Blue Ribbon factory. We drove north with a few cases of PBR beer tucked into the back!

FAA Pilot License

While in college the Air Force paid for 36 flight training hours for all pilot candidates. The Air Force didn't care if we earned a Private Pilot license or not. They just wanted us to have the flight experience prior to attending Flight School. I flew out of the Superior Municipal airport. We trained in Cessna 150s (single engine prop with 2 seats).

On one early dual training flight I entered downwind for a full stop landing. My instructor said, "I have the plane," and then he turned the fuel supply off killing the engine, then lifted nose so the propeller stopped spinning. He gave the plane back to me saying, "Your plane, land it." I completed the traffic pattern by flying downwind, turned base, then turned onto final and landed on the runway. After we stopped, I started the engine and we taxied in to parking. Airplanes get really quiet when the engine stops!

On a dual cross county training flight, we flew over Lake Nebagamon. I mentioned to my instructor that I lived "over there" pointing to the small Sunfish Lake to the southeast. I was curious when he started putting flight publications that were on the top of the glare shield down onto the floor. I soon had the answer when he said, "I have the plane." He pulled the power off and lifted the nose up which put the plane into a stall, then the plane nosed over down into a spin!

We were spinning fast! My first spin. I couldn't even focus on anything outside the plane, I'm thinking (hoping) this guy must know what he is doing? He recovers over Lake Nebagamon and heads in the direction of Sunfish Lake. He came out of the spin low enough that we had to climb up to clear the tall pine trees along Lake Nebagamon shoreline. I pointed out to the left at Sunfish Lake and my house. He banked left and descended to just over the water. We flew from the south over Sunfish Lake directly toward the house. The house was above us. It just so happens that my mom was standing in the yard watching and my dad was driving down the driveway returning from work. We crossed the lake then banked right, climbed, and flew back to Superior. Later that evening I called home from my Superior apartment. Mom was really upset and asked me if I was flying the plane.... she was not happy!

Many years later I've asked a few of my fellow ROTC cadets if he did this type of activity with them. "No," was their answer. So, I'm thinking, wondering, what was it about me that he would do these aggressive maneuvers with me as a student pilot?

I soloed in Cessna 6249R on 2 October 1968 after only 11 hours of dual instruction.

During my senior year, on 25 May, 1969, I flew solo to Siren (46 minutes) for my ROTC progress check and FAA Private Pilot check rides in Cessna 4319U. Siren was a small airport without a control tower. There was absolutely no activity on the field at the airport. When I arrived, there wasn't anyone around and the small terminal building door was locked. I sat down on a bench at the front of the building and waited.

A Cessna 150 entered the traffic pattern and landed. It taxied within 50 feet of me and shutdown. The pilot walked over to me with a map in hand, and asked me, "Where am I?" He was on a solo cross-country training flight out of the Twin Cities and was lost. I pointed to the Siren airport on his map. He said, "Thank you" and turned around, started the plane, and left. I wonder if he made it back to the Twin Cities.

The FAA check pilot arrived. The two check rides lasted one hour. I passed both! I was officially a pilot! I flew back to Superior (54 minutes).

A few days later, I checked out in a Cessna 172 (8542U) (4 seats). The cost to rent a Cessna 150/172 was

$12/$16 an hour! BTW, an instructor pilot cost was $3/hour.

I was commissioned a Second Lieutenant in the Air Force, on 4 June 1969! My orders were to attend pilot training flight school (class 71-02) at Randolph AFB, San Antonio, TX.

Mom, Dad, and I traveled to the Duluth Air Guard Base transportation section. The Air Force issued me airline tickets from Duluth to San Antonio on Northwest Orient Airlines. I thought it was impressive that my airline travel costs were actually being paid for! I had never before flown on a commercial airliner.

As a pilot, my first ever passengers were Billy and Larry. I flew them on 11 June 1969 for 50 minutes. I flew a number of friends during the following weeks.

I flew my Mom, Dad, and Sister (Bev) for 1 hour on 21 August 1969. The next day, 22 August, 1969, I departed for the Air Force. (I was in the Air Force for 30 years and seven days).

My senior year (1969) income was $2,068 and I paid $30.90 in Wisconsin Income Tax.

Chapter 2 - Air Force Training

Flight School

On 22 August 1969, my parents drove me to the Duluth airport. On the way, we stopped at the Poplar State Bank, along highway 2. I was totally broke. I borrowed $120 cash on a 90-day signature loan so I would have money to live on until I started to receive military pay. I also owed about $3,000 in college loans. I entered the Air Force with two suitcases, $120 of borrowed money, and college loans.

In those days' airport security was different. Back then folks weren't trying to kill us so there was no TSA, no metal detectors, and no x-ray machines. My parents walked with me out onto the Duluth aircraft parking ramp to the aircraft. We hugged and said farewell at the bottom of the aircraft loading stairs.

I flew from the Duluth Airport on a Northwest Orient Convair 580 twin engine turbo prop to the Minneapolis airport. I was 22 years old and this was my first ever airline flight.

At Minneapolis, I wandered through the airport terminal until I found my DC-9, jet engine flight to San

Antonio. These were my first airline flights! I wore my 2nd Lieutenant class-A uniform.

It was dark when I arrived in San Antonio, TX. I walked through the airport wondering how I would get to Randolph Air Force Base. I came upon a "Military" reception desk. I didn't know what that was but since it said "Military" I stopped.

I was told that a motor pool car from Randolph had come to pick up a high-ranking officer. But the officer didn't show up so I was given the car to travel to Randolph. The driver was about as old as I was. I remember how strange it was when he took my two bags from me and carried them to the car. And then he opened the back door for me! (Welcome to the world of the military.)

I remember seeing Randolph AFB off in the distance from the highway 35, and the driver pointed out the Taj Tower (all white-water tower and Base Headquarters) sitting approximately in the center of the Base. I checked in at the billeting office and was assigned to a building and room.

My first room was on the second floor of an old WWII building. The apartment had two small bedrooms, a shared bathroom, and a small front room/kitchen. My first roommate had almost completed flight school and was scheduled to graduate within a week. I don't remember his name. I recall getting up in the dark one night to go to the bathroom. I stepped on a number of cockroaches in my bare feet. My first ever cockroaches, but not my last!

The next day I met another member of my flight school class. He drove me into San Antonio where I bought my first new car. It was a tan 1969 VW bug with an air conditioner! It cost $2,300. I now had school loans and car payments!

More than half of my pilot training class members were air force academy graduates. They knew each other and obviously were used to being in a military type organization. Compared to the academy graduates, the ROTC training did little to prepare me for the military experience. Also, it was my first time away from my home turf, friends, and family! I admit that it was a very difficult time for me.

We spent the first week in-processing. A couple of incidents are memorable:

One event was flight physicals. They were conducted in the basement of the old Randolph Base hospital. At one point a number of us were seated in a circle with our shoes and socks off. A Tech Sergeant was taking our footprints with ink onto paper. We were a bunch of cocky smart-ass 2nd Lieutenants making jokes as to what the prints were for. I asked him, "So why the foot prints?" He looks up and says, "You don't know why they are taken?" I said, "No, what are they used for?" He said, "Follow me," and he led us down the hall to the hospital morgue.

In we go and he pulls a body tray out from the morgue wall. There is a body in a semi clear plastic bag. It was a body of a pilot. I could see a burned body, flight suit, and boots. The Tech Sergeant points to the body and says, "See, when a plane crashes and burns, the feet are protected by the leather boots and are the last to burn." The foot prints were for identification! Reality check! One week in the Air Force and see my first body! (Many years later we had come a long way when DNA samples were taken and used for more accurate body identification.)

The other event occurred during the flight physical when a female airman asked me to sit on a table and put my back against the wall. Since I had been through this during the physical in Duluth, I knew what the purpose was for the test. It was to check my sitting height. If sitting height was too tall then no flight school. I wasn't concerned since I understood that the requirement height limit had been increased and I was within the limits.

I noted she started to write something down. I asked her what she was writing. She commented, "Your sitting height was too tall to be a pilot!" What! I got pretty upset, extremely verbal, and I made a big fuss. A Tech Sergeant came over and asked, "What is the problem?" I said that, "I had passed the exam in college, and was now in the Air Force at flight school and suddenly my sitting height was too tall to be a pilot." He said to try again. So, I sat back against the wall. He takes a look and said that my sitting height looked, "Close enough."

Wow, "Close enough." Thinking back, had I not asked what she was recording, and had not made the big fuss, I would have been reassigned from flight school to do whatever.

(In reality, when I sat in the T38 jet trainer the top of my head was above the ejection seat canopy breaker spike. Meaning, if I had to eject my head would have hit the canopy before the canopy breaker spike.)

I wish now I could thank the Tech Sergeant who gave me the "close enough" call. Apparently, my sitting height must have been too tall but he looked the other way. I spent 16 years flying Air Force aircraft! (BTW, sitting height wasn't an issue in the B52.)

About two weeks after arriving at Randolph my mom calls and tells me I received an official looking letter from the government. I asked her to open it. She did, it was a draft notice for me. So, they wanted to draft me about 2 months after graduating from college. I could have been pushing swamp water in Vietnam if I wasn't already in the military. My freshman decision to remain in ROTC 4 years earlier was a good one.

Mom asked me, "What should I do with it?" I told her, "Don't do anything, we'll let the draft board figure it out on their own." I wish now that I had told her to keep the draft notice! I was to be drafted 2 weeks after I was already in the military!

Three weeks after I bought my first new car, the 1969 VW bug ($2,300), I decide to drive down to Laredo, TX, and visit a college buddy attending pilot training there. On Friday afternoon I headed toward Laredo. I planned to return on Sunday evening. As I drove south near Cotulla, I came upon policeman along the highway who flagged me down. I rolled down the window. I'm asked for my driver's license and proof of insurance. They looked ok, but then he tapped on my windshield with his nightstick and says, "Your vehicle safety inspection sticker was out of date." I said, "That couldn't be possible, I had just bought the car three weeks ago." He said, "New car safety stickers are only good for two weeks." He then said, "Follow me to the Cotulla police station." I followed him and we parked in front of the station.

I walked through the front door of the police station it looked like the Mayberry jail house (as in Barney Fife in Mayberry)! There was a low fence across the room with swing gates in the middle. Two desks faced each other, one on each side of the fence. Further back, were two jail cells. He tapped on the chair with his night stick and said, "Sit here." He went through the gates and sat down on the other side of the desk facing me. He picked up the phone and dialed a number. He asks if Billy-Bob is there. No? He dialed two more numbers and asks the same question. He can't find Billy-Bob, the local judge. He put the phone down and drums his fingernails awhile on the desk. Then he said to me, "I'd hate to have to put you in jail all weekend." With that he got my attention, jail, all weekend, for a safety inspection! He then said, "Can you get the vehicle safety inspection done as soon as you can?" I said, "Yes I can, as soon as I get to Laredo!" He let me go. I drove into Laredo and the first thing I did was to have my three-week-old vehicle inspected. My take is that he, the town, and the hanging judge used the highway to generate funds for the town. But the judge was out of town all weekend. Apparently, this policeman had a conscience.

The next weekend, on weekend four, I drove to Del Rio, TX, to visit Scott. He was attending flight school there. I stopped in Uvalde, TX, for a late morning breakfast. I'm sitting at the counter reading the menu. I see, "Chicken Fried Steak and Eggs" on the menu. I'm thinking what the heck is that? Chicken? Steak? Chicken Fried Steak? What? Being adventurous I ordered it. My first ever Chicken Fried Steak and Eggs, October 1969! It was huge! It hung over the sides of the plate and was covered with gravy. It continues to be my favorite breakfast!

It didn't take long for me to realize that it was a very big world and I was very new to this world. I grew up a "Jack Pine Savage" (Rough & backwoods upbringing) in the woods of northern Wisconsin with Norwegians, Swedes, Finlander's, and Germans. The only Mexican I knew was on the Saturday morning TV Bugs Bunny cartoons. His name was Speedy Gonzales!

My second roommate at Randolph was classmate Rick Gonzales. I was moved down to the first floor of the old WWII building. Rick was a Hispanic ROTC graduate from San Jose, California. Rick was a nice man. He was the only Hispanic in my class. After one-month Rick "self-eliminated" from flight school, in other words, he quit. I don't understand it to this day.

One day Rick said, "Let's go down town for a Mexican dinner." We ended up at a restaurant patio eating Tex-Mex and engaged in a discussion with four Hispanic men at the next table. It turned out that they were employed at Kelly AFB, San Antonio, where they drove forklifts in the supply warehouse. Ok, I drove forklifts on the docks in Duluth so I could speak forklift language with them.

They invited us to follow them to a nightclub. We pulled up in front of the night club. A large lettered sign arched over the front door that read, "Gonzales Cabaret."

The six of us are seated and talking. I see a beautiful girl seated alone so I go over and asked her to dance. We dance a couple dances. After dancing, we sat at her table and talked for awhile. Then she said something with a heavy Hispanic accent that didn't at first totally register to me. I said what? She repeated and said again, "You're very brave!" I asked her why she would say that. She said, "You are the first gringo to be here in 4 years."

With her comment, I thought back through the evening. When we first arrived Rick and our new four friends sat at a picnic bench type table, I'm at the right end next to the wall. I looked to my right and there about two feet away was a security guy leaning against the wall just next to me. I recall thinking how big the holstered .45 on his hip looked. I thought nothing of it at the time.

Later, I went to the bathroom. I'm using the urinal, I thought it strange that one of my new Hispanic friends followed me into the bathroom. I thought it curious that he just leaned against a wall while I used the urinal. He then walked out after me.

It was then that I realized they were protecting me. I talked with the girl for a few more minutes then went back to the group table. A few minutes later I told Rick we needed to leave which we did. It was all a bit too much like the Marty Robbins' ballad, El Paso.

When Rick self-eliminated, I moved from the old WWII building to a brand-new building. The floorplan was different. We each had our own bedroom and bathroom. There was a narrow-shared kitchen between our rooms. (BTW, this was so long ago that this "new" 1969 building has since been torn down and replaced.)

My last roommate was an Iranian. There were a number of Iranians attending flight training at Randolph with us. Each class had at least two Iranian students in it.

Someone told me that if they were doing poorly that an Iranian Colonel located across town at Lackland AFB would beat them. One day my roommate told me that he did very badly during a flight lesson that day. He told me that he pretended not to understand what the instructor wanted him to do. I wonder what his fate was after the revolution.

There were also South Vietnamese pilots' training at Randolph. Their call sign was Baby. They had their own program and we didn't have much interaction with them.

Flight school was a year of 53 intense weeks. I was new to the military way of life, flight training was a pressure packed environment, and it was my first time away from home. I recall going to the base post office and checking my mail box daily. Nothing! How sad that was?

There was a bar in the basement of the Randolph Officers club called the Auger Inn. Auger Inn is a reference to crashing an aircraft. It was odd to dance at the club on Friday nights in a flight suit and with flight (combat) boots on.

During the year, my class lost a few of students due to self-elimination. Thankfully, none were injured or died while training. We did, however, have one die just a few days before graduation in a car crash. He had a fast car.

I entered the Air Force broke (with $120 loaned money). Back then the AF paid us twice a month by check. We went to the finance pay window and signed for our checks. I paid my bills, expenses, and then put $12 dollars from each payday into a Wells Fargo saving account. That's right. I saved $24 a month! Wow! One of my life guiding philosophies is to live below my means. I was doing it, but just barely.

Here is what my 1969 Air Force monthly pay was:

Base Pay	$417.60	Fed Inc Tax	$86.38
Incentive Pay	$100.00	Soc Sec	$20.04
Subsistence	$ 47.88	Allotments	$80.19
		SGLI Ins	$ 3.00
	$565.48		$189.61
	Net Monthly: $375.87		

During flight school we trained on three different aircraft, the T41 (a supped-up Cessna 172), the T37 (first twin engine jet), and a T38 (supersonic twin engine jet).

My class (71-02) had 80 students. We were divided into flights A & B. I was in B flight. The days were split with one flight attending academics while the other flight was flying.

We were bused from Randolph to Stinson airport in

San Antonio for the T41 training. The instructors were civilian contract pilots. We flew the T41 for about 12 hours.

The T37 and T38 flights were flown out of Randolph; the instructors were Air Force pilots. We flew the T37 for about 60 hours and the T38 for 90 hours.

Student Pilot Section B

In 1969, at flight school, I wore my first helmet. The helmet was a hard-white shell with ear cups inside. There was an oxygen mask with a bayonet receiver on each side. The receivers were used to attach the mask to the helmet. There were two visors that could be lowered down against the oxygen mask. One visor was clear, the other dark.

Not all heads come in the same shape or size. To make the helmet fit the head foam rubber pads were stuck around inside the helmet to pad the head

The oxygen mask needed to be placed firmly against the face to form a seal. This caused the helmet to press against the top of the head. The idea was if the pads were correctly placed then the pressure would be distributed equally over the head. If not correct, then an uncomfortable "hot spot" would occur. Placement of the pads was hit and miss.

My first helmet didn't fit, a miss! My first few one-hour long T37 flights were hell. After just a few minutes a hot spot developed which felt like being burned. I couldn't think about anything but the pain. Post flight the pads were adjusted, moved around, over a number of times before the hot spot site-specific pain finally went away.

Years later the form fitted helmet appeared. A flexible cloth skull cap was placed over my head. Then a metal cover placed over that. A chemical mixture was poured between the two that foamed up creating a mold of my head! The foam mold was then trimmed and placed in the helmet. The fit was great! Helmet problem fixed.

Also, my Oxygen mask didn't fit correctly. I couldn't get a good seal with Air Force issued masks; maybe my nose was too big, or whatever. I needed a custom-made mask. So, I was sent to the dentist office where a mold was made of my face. As I was laying back in the dentist chair, straws were inserted into my nose so I could breath. Cotton pads were put over my eyes. Plaster was then pored over my face and mouth to make a mold. I laid there for about 20 minutes, breathing slowly and carefully through the nose straws, until the mold set.

The mold was sent off base where a custom mask was made. The custom mask fit perfectly. Every couple of years I needed a new custom-made mask. I wonder where the mold is now.

Randolph AFB is located Northeast of San Antonio and the T37 training areas are to the west. Each training area was a large box of airspace up to 25,000 feet. We were assigned an area to practice in. Fixed routing & altitudes to/from these areas were published. The routing to the areas was around the south side of the city at 14,000 feet, return was around the north side.

First Jet Trainer, T37 Twin Engine

During one of my first T37 solo flights I was in a training area west of San Antonio practicing aerobatics. I was at the top of a loop (upside down) and heard a loud pop noise in my headset. What? I lost my radio! What to do? I was com out (no radios)!

I saw another T37 on the published return routing to Randolph and pulled up next to it. There was an instructor pilot and a student pilot in it. I gave them the visual hand gestures "lost my radio." They acknowledged and slide back into formation off my wing. They made the necessary radio calls as we flew the routing back to Randolph for landing. I rocked my wings on final indicating I had no functioning radio (com out) and tower controllers gave me a steady green "cleared to land" light.

On my second T37 solo flight I lost the right engine. I entered the overhead visual traffic pattern for a pitchout intending to accomplish a touch and go landing. I couldn't pull the two throttles back as I started my descending turn onto final (align with the runway).

It was somewhat confusing. What was to be a touch and go turned into a full stop. On the runway, I called the Runway Supervisor Unit and announced I was aborting the touch and go to make a full stop. He asked, "Why?" I said, "I can't control an engine." I turned off the runway, stopped, and raised the canopy to the full up position. Dense smoke was rolling up from both the front and rear end of the right engine. This was close enough that I could reach over and put my hand into the smoke.

I could not pull the right throttle back to idle so I shutdown the engine with the Fuel Shutoff T-Handle which cut the fuel off. A fire crash truck pulled up directly in front of me and swung a "huge" nozzle out in front of me. It was about 3 feet in front of the aircraft and looked really big! I had opened up the canopy. I was hoping that the firefighters wouldn't need to turn it on!

I rode the crew shuttle van back to my flight training room. As soon as I walked through the door the scheduler grabbed me by the arm and assigned me another aircraft tail number (assigned me to an aircraft) and told me to go fly. I guess they didn't want to give me any time to think about what had just happened. In about 45 minutes, I was back in the air!

I flew back to Wisconsin for the Christmas break. Had a great time visiting with friends and family.

The third aircraft flown was a T38. We flew it for a total of about 90 hours. It was a twin-engine supersonic trainer. During training we only went supersonic once.

Christmas Break. Note: no Wings on Uniform

On my first orientation flight (called the dollar ride) the instructor pilot asked me what I wanted to do. I said, "Let's do a maximum rate aileron roll." "Ok," he says and he then asked, "Are you ready?" I said, "Yes," putting my elbows against the cockpit sides. He yanked the stick to the left. My helmeted head slammed against the canopy on the right side as the aircraft snapped left, then he stopped and my head slammed against the canopy on the left side. Totally rang my bell big time! The instructor just chuckled. He knew what would happen.

We didn't see anything out of the T38 for the first number of flights. They were instrument training flights that focused on how to fly the plane using just the cockpit flight instruments. We were in the rear seat for this training and just after takeoff we pulled a curtain forward covering the canopy so we couldn't see outside.

Eventually, my class had its first solo T38 cross country flight from Randolph to Blytheville AFB, AK. We were a stream of student pilots flying solo aircraft. I had filed for 30,000 feet. What did I know? So, after takeoff I climbed to my filed altitude of 30,000 rather than the ATC cleared 20,000 feet.

I arrived at Blytheville with more fuel then my others so I did a few approaches. Then I landed. I screwed up big time! I could have been in someone else's airspace (like an airliner!) I should have leveled at 20,000 feet. Air Traffic Control information was different back then. At that time, ATC could only see my location on their radar. Today

ATC can see aircraft location and aircraft altitude readout on their radar.

I flew a dual (with an instructor) cross-country training flight to the west. Our refueling stop was at

The G-Suit was good to 9 G's

Holloman, AFB. The temperature was really hot. Takeoff computation were so critical that the instructor wanted to make the takeoff. We landed at Nellis, AFB, Nevada.

That evening we went into the Officers Club. There were topless dancers in the stag bar! WOW, topless dancers! Never saw that before! Don't worry if your offended, topless dancers in Air Force clubs ended years ago. Darn!

I graduated from Flight School on 1 September 1970 and was assigned to B52G models at the 744th Bomb Squadron located at Beale AFB, CA. Aircraft assignments were made by class rankings. I struggled with both academics and flight training. I was ranked really low in the class of 80 students.

Most of the instructor pilots at flight school were fighter pilots who had nothing good to say about multiengine aircraft. The culture at that time was fighter pilot culture. Back then I was actually embarrassed to be assigned to a B52, a multi engine plane.

My class went on to fly various aircraft. I soon learned that flying the B52 was a daunting task and something of which to be proud. I actually became highly pissed with my pilot training instructors and their tactical culture. I served 16 proud years associated with B52s in the Strategic Air Command.

My mom, dad, and sister (Bev) flew into San Antonio to attend my flight school graduation ceremony. We had a great time. We toured around San Antonio and the base. I showed them the aircraft and had them sit in the T38! Dad even flew the T37 simulator.

After my graduation mom and Bev flew back to Illinois. Dad stayed and we drove my 1969 VW north. I had some leave time (vacation) to take before I needed to report to Beale, AFB. We took turns and drove all night arriving into Bull Shoals, AR.

We stopped at a restaurant for breakfast. We asked the owner if they knew my cousin Butch. We understood that Butch lived on the lake and owned/ran a boat launch marina. No was the restaurant owner's response, he never heard of the name. So, dad and I left, intending to continue our drive north.

Heading north, we drove around the lake and saw my cousin's marina sign so we turned in. Butch is on a dock. As we walked toward him, at first, he didn't recognize us so that was my cue to screw with him. I say we are planning a fishing party for 200 folks and can he take care of us. His eyes got big but then he figured out who we were and we all had a good laugh.

About that time a motor boat pulls up. It is the restaurant owner who earlier didn't know Butch! He says, "Great you found him." What the heck? Who is this jerk? Anyway, we had a nice but short reunion with Butch. We then continued north. I dropped dad off in Lombard, IL, where my mom was visiting my Nana. I continued north to Carpentersville and visited Uncle/Aunt/Cousins, then continued to northern Wisconsin.

I had a great time visiting family and friends in and around the Town of Highland. Then I left for my first drive across this country. This is one big beautiful country with lots of first-time discoveries for me! I drove west on highway 80. One dark night I stopped in at a motel. In the morning I left the motel, looked around and saw all the mountains around me…. wow! This flatland kid had never seen mountains before!

Then along the Salt Lake I actually stopped and tasted the water! Yes, it's salty. As I climbed up the Rocky

Mountains my VW began to lose power. I forget which town I stopped in but I stopped at a VW dealer and told them of my problem. They tweaked fuel/air mixture on the carburetor and the VW ran fine from then on. It was the higher altitude and thinner air was the problem. I departed Highway 80 onto 20 and drove down to Marysville, CA. The mountains, the trees, and the views! It was a beautiful drive!

Beale AFB

I arrived at Beale on 28 September 1970. Soon after I arrived at Beale a Strategic Air Command Operational Readiness Inspection began. I was not combat crew ready so the Squadron Operations Officer told me to get lost until the inspection was over. I got lost.

Every month the Bomb Squadron Commander would have a Commanders Call. Everyone in the squadron attended. Information and issues were talked about and announcements were made. It was held at the Beale Officers Club. I'm a gold bar 2nd Lt wearing my flight suit with my pilot wings on it.

Have you ever watched old WWII movies depicting a mass briefing of bomber crews? I have. That was it except now I was in the movie! As I walked past rows of B52 flight suited crews I suddenly realized I was really in a combat squadron! It was the Cold War years. These crewmembers sat nuclear (cold war) alert, they had flown combat sorties in South East Asia, where some flew hundreds of combat missions, and they were flying 24-hour airborne alert sorties with nuclear weapons onboard.

Some of these crewmembers had flown in bombers during WWII and during the Korean War. Some had thousands of hours in the B52. It seemed that most were Lieutenant Colonels. I'm a gold bar 2nd Lt and hadn't even been in a B52 yet!

To become a B52 combat crewmember I still needed to attend Nuclear Weapons Shool, Air Force Survival School, B52 Combat Crew Training School, and then complete B52F to G model transition flight training. Finally, I needed to pass a Wing Emergency War Order Certification Briefing.

I flew into Fort Worth, TX, and attended the SAC Air Weapons Delivery School located at Carswell AFB, 8-19 November 1970. I studied about nuclear weapons! I learned how the safety systems worked, about weapon settings, how they were armed, and the delivery tactics used. It all was classified so no materials could leave the classroom. When was the last time you studied nuclear weapons?

My first weapon. Years earlier, on my 13th birthday, my folks drove me up to Duluth, MN. With a bag full of S&H Green Stamp books, I selected a .22 single shot rifle. We walked down the street with me carrying the rifle. Now I was training about nuclear weapons. (BTW, I don't have any Nukes, but I still have the rifle.)

Four Nucs were carried in the forward Bomb Bay

One day I was having lunch in the Carswell Officers Club. Three Colonels came up to meet me and to shake my hand. They had never seen a 2nd Lt (gold bar) with pilot wings before! Nuclear training complete I flew back to Beale AFB.

Survival School

Next, from Beale I drove my VW cross country north to Spokane, WA, to attend Basic Survival Training School at Fairchild AFB from 22 Nov to 7 Dec, 1970. This was the time of Hippies. Young long-haired people were hitchhiking everywhere all across the country. I never picked up any, however, I came upon an older man in Oregon and gave him ride.

We had a long talk as I drove. This man was very respectful, especially when he heard I was a 2nd Lt! He said he had been a marine during WWII and fought in the Pacific Islands. He didn't talk about his WWII Pacific Island battle experiences. However, he told me about Marine training, said that once his Marine Company was put into a muddy circle, and told that the last three marines remaining in the circle would get a 3-day pass. He said he was one who got the leave. I believed him. I dropped him off and offered him $20.

Survival training was a real eye opener to me. There were two parts to survival school, Basic Survival training, and then Prisoner of War (POW) Survival. POW survival was divided into the torture/isolation experienced in South East Asia (SEA), and Mass Cold War communist prisoner camps. Given what I know now, I hope you have a safe space to hide in?

During Basic survival we attended many academic survival classes, then were bused into the Spokane Mountains near Cusic. It was November so we used snowshoes to walk into the mountain forest where we lived in the snow. There were about 15 students in each camp. We learned how to build a fire, how to make shelters out of parachutes and branches, how to catch food, and how to make rescue signals. I fired off signal flares, gyro rockets, and practiced with the signal mirror. We built snares, talked about water, and emergency first aid.

Snowshoes into the Mountain Forest Camp

The first night two of us used a parachute to make a lean-to shelter. Other nights we made a shelter by snapping two rubber rain ponchos together. I chopped wood for our fire. Ok, I'm a jack pine savage from Wisconsin so I'm used to snow and cold, but there were kids attending who have never crossed the ditch before. There was a kid shaking with cold squatting in the snow. I shoved him and made him get up and move around. I told him he could die from the cold so get moving. Through the years people did die during this training. The key to survival is to never give up. He had never experienced cold and snow and just gave up.

Survival Training Camp

I have to admit I did get upset on one occasion. One night after a few days in the field I was slipping into my sleeping bag. Two of us had snapped our rubber rain poncho's together forming a rubber tent. I was taking very deliberate actions and I finally get all settled in. I pulled the sleeping bag up and around my face, done. I look up and see the poncho hood hanging down inside the tent. Oh man, I thought, if I don't push it up and outside then water may eventually leak down into the hood and begin to drip on me getting me wet. So, I loosen up the sleeping bag and reach up to push the poncho hood up and out. Well, the poncho head was already full of ice-cold water! Ice cold water poured down my arm into my arm pit and into my bag! I almost screamed!

After we returned from the mountains and basic survival, we attended a number of classes relating to Prisoner of War (POW) training and issues. We discussed Escape and Evasion techniques. That is, when and how to move to avoid being captured.

We discussed the Code of Conduct. We knew that US POWs in Southeast Asia (SEA) were being tortured and treated horribly. They were being used for propaganda. They were being tortured in an effort to get them to tell the world how good they were being treated. Just give that a moment's thought. Finally, I recall being told that if I became a POW, and if I survived, then I would be held accountable for my actions while in captivity.

POW training started at night by crawling on our bellies through a simulated battle field. It was an obstacle course where we practiced escape and evasion techniques. I crawled through the mud around, over, and under obstacles. There were trip wires everywhere connected to flares to avoid. We wore our green fatigue uniforms. This crawl was the transition from escape and evading to being captured. As I went, I could see the ghost like shapes of other evaders moving around me. Eventually I got to the end and slid into a ditch where I lay for several minutes. Looking back, I watched flares burning behind me. Others dropped into the ditch. I didn't know who they were.

In front I could see other evaders leave the ditch, walk forward, and be captured. Did I really want to keep going and get captured? I told myself that this was only training. Even so, I was a little apprehensive. I had carried several small plastic baby bottles of water with me. I drank the water, stood up, walked forward, and was captured.

A bag was immediately pulled over my head, I couldn't see a thing. Not seeing is a bit unsettling since I couldn't see what was happening nor see what was about to happen. The guards put my right hand on the "war criminals" shoulder in front of me. We were a large group assembled into a circle. We walked around in a circle for awhile. Now, my guess is they were waiting for all the class to be "captured."

Eventually, I was led, still blind from the bag over head, inside a building and shoved into a closet like cage. Hooded, I was ordered to remain standing. The box had solid wood walls and door. It was about 3 by 3 feet in size, no problem there. However, I'm 6'3"and it was about 5' 8" tall meaning that to stand hooded I had to bend way over. I don't know how long I was in the cage.

Eventually I sat down but kept listening for footsteps because the guards would sneak around and occasionally jerk a door open to check on a "war criminal." Over time I could hear other "war criminals" being removed and returned to the cages.

At some point I was taken, still hooded, to another location and sat down on the floor with my back against a wall. I took the opportunity to sneak a peek for a moment from under my hood. There were a number of us sitting shoulder to shoulder on the floor. I saw a camera up in a corner. I wondered was that for them to watch us, or was it there to watch what our "guards" were doing to us? Behind and above us was a wall of many small doors.

I soon learned what the doors were for. I was stuffed into a very small cage. I was on my knees which were tucked up under me against my chest and on my elbows. My head was bent down toward my knees. Any available room was then eliminated as wooden slats were pushed in around me to further tighten the cage against me. I couldn't move. This was not a comfortable thing. Someone claustrophobic would have a real difficult time with this.

They either played the sound of someone screaming or someone was actually screaming. I couldn't tell if it was real. Are you claustrophobic? I don't know how long I was in the box.

At some point I was taken for a one on one interrogation. The hood came off. The room was a replica of a North Vietnamese prison camp interrogation room.

As I entered the room, I was relieved a bit when I noted there wasn't a hook sticking down from the center of the ceiling. No hanging from the ceiling by my arms tied behind me! Nor was there a steel bar on the floor so ropes wouldn't be used to pull my arms over my shoulders! The ropes! (With arms behind their backs, ropes would be wrapped tightly around the arms until the elbows touched. The ankles were attached to the bar on the floor. A rope from the arms were placed over a shoulder to the bar. The arms were then forcibly lifted and pulled/pushed up dislocating the shoulders.)

There was a stark desk/chair, a light bulb, and a stool. The NVA interrogator taunted me about my family, my physical traits, and my country. I was a war criminal, not a POW, nobody knew I was alive, and finally, I would never go home.

I was made to stand with my back against the wall, put my hands up and over my shoulders against the wall. Then I had to move my feet away from the wall. This was very uncomfortable arched over backward and my entire body soon shook uncontrollably.

Then my interrogator accused me of carrying contraband food/candy items in my uniform jacket pockets. I denied it. I said I had nothing on me. "Ok, prove it." He said, "Take off your fatigue jacket, throw it on the floor and stomp on it, and convince me."

I did what he asked and then he said, "Thanks. I now have a video of you, a US pilot, rejecting his service and country by stomping on your uniform." I learned I could be manipulated. The important lesson I learned was not to believe anything I saw, heard, or was told.

At some point the POW training situation transformed from North Vietnam isolation/torture to a Communist prisoner of war mass prison camp. I was moved from isolation into a mass prisoner camp with all the other prisoners.

Remember at the time in my service we were fighting two wars, SEA (Vietnam) War and the Cold War (Soviet Union). It was night. There were tall barbed wire fences with guard towers and spot lights. There was a large concrete bunker without doors. It was really cold. In the bunker we were given lectures about the evils of capitalism and the greatness of communism.

A prisoner actually escaped and got through the wire but was quickly captured. He was put into a shallow cage dug into the ground. A steel grading pushed down against him.

It was snowing, I was really cold and was very hungry. Eventually we were lined up for a meal. The meal was being cooked in a garbage can over a wood fire. We were walked past it. Each prisoner received a bare hand full of cooked rice with sardines. It was great! Have you ever licked rice and sardines off your hand? I have.

After Survival School training I stopped sending event notices (graduations/promotions/moves) about me to be published in my hometown newspapers. During the SEA war enemy friendlies (Americans?) that lived in the US would search local newspapers for articles and information on the captives the North Vietnamese held. The NVA would use the information about you against you (Me, if captured). As a result, I never sent anything to be published in the media.

I learned a number of things as a result of this training. I decided that I didn't want to ever become a prisoner. It would be really uncomfortable when someone totally controls your food, comfort (hot/cold), bed/floor, shackles, and level/length of pain suffered. I learned that when you are a prisoner to never believe anything you see or hear. It's all done for a reason. I've known 6-year SEA POW's. I wonder how they survived and wondered if I could.

(20 years later I actually reverted to my POW training at Offutt AFB. In 1990, an army team was sent to Offutt to administer lie detector tests to Special Access Program officers. I didn't believe anything I was told, saw, or heard. The event is described later.)

Please read the book "Honor Bound, American POWs in SEA, 1961-1973, by Rochester & Kiley." It describes the total hell our POWs lived through. You won't believe what you read.

One Survival School laugh! One of my pilot flight school classmates (Henry) attended Survival Training in the same class. One day we were driving in my VW Bug through the Spokane streets. It was snowy/icy and I'm doing about 30 mph. Now Henry is a good old boy from near Tyler, Texas. A place not known for snow. I'm a jack pine savage from the snowy woods of Northern Wisconsin. We did this maneuver from time-to-time just because - we could! I pushed the clutch in, pulled up on the emergency brake, and turned the steering wheel. We did a quick 360-degree spin. I released the brake, let the clutch out, and just kept on driving down the street. I looked over at Sham (his nick name). His eyes were about the size of soft balls and his chin was on his chest. He then said a few unrepeatable words to me.

Combat Crew Training School

Initial B52F Combat Crew Training School was accomplished at Castle AFB, Merced, CA. I was assigned to a B52 initial class beginning 4 January 1971. It was 4 months long. I graduated 26 April. My training crew was unusual since there wasn't an aircraft commander (left seat Pilot) assigned to it, only three copilots (right seat). This was a good deal for us copilots because we shared the training that would have been the aircraft commanders among the three of us. I had the opportunity to make takeoffs, air refueling, and landings, that I wouldn't have had if there had been an aircraft commander on the training crew.

The B52F model was used for training. Castle was the only location that flew the F-model. After graduation each crewmember accomplished B52D, G, or H model difference training at their assigned unit.

I was selected for promotion to First Lieutenant while attending Combat Crew Training.

There were so many crewmembers attending initial B52F training that there wasn't room for all the students to live on Castle AFB. Motels around the area were contracted. I lived in the Pine Tree Motel for 4 months. It wasn't fancy. It was a one floor building situated in a U shape. I watched the initial series of Star Trek while I studied!

Back then, there were no computers, iPads, or iPhones. My first B52 pilot cockpit training consisted of two of us copilots sitting on two chairs facing a large picture of the B52 cockpit on the wall. We read the B52F checklist to learn the cockpit checklist flow and learned switch location by pointing at each switch. From there we moved to the non-motion simulator. The cockpit instruments operated while we practiced the checklist and flew simulator training sorties.

Academic classes covered all the various B52 aircraft systems. Large mockups of each aircraft system were used by instructors to demonstrate how the switches moved and how the systems functioned. We covered electrical, pneumatics, oxygen, flight controls, and fuel systems, to name a few.

Pine Tree Motel and my New VW Bug

Finally, we moved to the flight line and met our four flight instructors. Major D was our instructor pilot (He looked like Mr. Magu). My student training crew consisted of three copilots, a radar navigator, navigator, an electronic warfare officer, and a gunner.

Training flights occurred on a three-day cycle: A planning day, the flight day, and then a debriefing day.

The 10/11-hour training flights were extensively planned. The navigator produced a Form 200 with flight plan details. It contained headings, turn point coordinates, and leg times for the entire flight. The F200 was given to a copilot who calculated fuel used for each leg. Fuel burn rate varied with gross weight, altitude, and speed. The result was a fuel flow. The leg method resulted in the total fuel burn for that leg. A fuel use graph was produced. The copilot also calculated the weight and balance for the aircraft. Takeoff data was calculated based on gross weight, outside air temperature and pressure altitude.

The planning ended with a detailed lengthy mission briefing by each specialty, then as a crew.

The second day we flew the training sortie. A typical flight included: preflight, taxi, takeoff, climb, a 1 ½ hour celestial navigation leg (sextant shots, sun/stars), rendezvous with a KC135 and air refueling, descent into a terrain avoidance low level, three nuclear bomb runs, climb out, flight back to Castle AFB, descent, instrument approach, and multiple touch and go landings, then a full stop.

The third day we did a detailed debrief discussing the good, bad, and ugly that had occurred during flight. This debriefing was done first as a crew then by aircrew specialty.

During the training flight we couldn't stop and talk about what happened. I learned to create a list of notes about occurring events that I didn't understand or went badly to research later. Over time, as I learned, the list got shorter and shorter.

The B52 is a big plane. Wingtip to wingtip was 185 feet. Nose to tail 160 feet. Empty weight was about 183,000 lbs. People, fuel, and munitions could be loaded to a maximum gross weight of 488,000 lbs. The wings were very flexible. There are 12 fuel tanks. With a large fuel load the wings drooped down. There is a tip gear near each wingtip. As the plane accelerated during takeoff roll the wing tips started to lift at around 70 knots. A max gross weight takeoff occurred at 163 knots (178 mph) after an 8 - 9,000 feet of takeoff roll.

A cold war nuclear tactic was to fly low-level using terrain masking to avoid enemy radars. I recall my first training flight low level. The terrain avoidance (TA) radar equipment could guide the plane down to 200 feet crossing the ground but we usually practiced around 4-500 feet. You could dial in the altitude needed. For this first flight one of the other copilots was in the copilot seat. I sat toward the back of the upper deck compartment.

On the F-model, there were small windows that looked out over the wings. I couldn't believe what I was seeing. The wings flexed up and down while the engine pods moving forward and back. How was that possible? I stopped looking out the window!

I recall the first time I rolled onto a visual final approach to land the B52F. I commented over the intercom on how small the runway looked. My last landing was in a T38, really small jet trainer compared to this B52F with a wing span of 185 feet! Maj D answered, "Just land your seat on the runway centerline and the plane will land around you." I did, and it did.

Chapter 3 - Combat Squadron, Alert, & War

Combat Crew Training complete, I drove up the Sacramento valley to Beale AFB. I was checked out in the B52F. Beale flew B52G model bombers. I needed to accomplish difference training at Beale. The F and G model aircraft looked pretty much alike externally but they weren't. There were significant differences in all the aircraft systems. I completed the G model training, passing yet another check ride.

I lived in a serious world. We were fighting two wars: The Cold War (Soviet Union) and South East Asia (Vietnam) war. The Strategic Air Command (SAC) was on nuclear war footing. Nuclear weapons' loaded bombers at squadrons across the country were tasked to takeoff within 15 minutes of notification, day or night. Soviet missile flight time to the US was 30 minutes. Also, SAC crews were deployed to the South East Asia (Vietnam) war. From Guam and Utapoa, Thailand, they were dropping iron bombs in Laos, Cambodia, and Vietnam.

The B52 Squadron Commander was a Lieutenant Colonel, the Operations Officer was a Major. There were twenty combat ready crews in the squadron. With spares there were about 150 crewmembers in the squadron.

When I arrived in 1970 the military was mostly a male world. There were no women in the Bomb Squadron.

Around 1973, women began to appear in the Wing. There was a married couple (Lieutenants) that were intelligence briefers. Later more women appeared in the maintenance community on the flight line.

There was always something new to learn. I was a new copilot just out of initial B52 training on a night training flight. The Squadron Operations Officer (Instructor Pilot) was in the left seat, I'm in the right. It's my turn to make the landing.

The B52 has a unique characteristic. It fly's nose low meaning the front landing gear are lower than the aft gear. During a landing, as the airspeed bleeds off the nose of

488,000 lb. Gross Weight, 185 feet Wing Tip to Tip

the aircraft and forward gear rotates up and the aft gear down. The ideal landing occurs when the aft landing gear touches down at the time the forward gear is about half wheel height above the runway.

As the plane slows the aircraft attitude changes. There is a simultaneous touchdown speed at which all gear touch down at the same time. Above simultaneous touchdown speed, the forward gear touches down first, slower than it, the aft gear touches down first.

If the forward gear hits first the nose bounces up, then the aft gear hits bouncing the plane back up into the air. Without correct control inputs the plane will porpoise with another forward gear, then aft, and the plane goes airborne even higher. Each porpoise gets bigger and more violent.

All the while airspeed is decreasing. I landed on the forward gear and we bounced airborne. I hit the runway a second time and we went airborne a second time higher. I had never experienced this situation before. At the top of the second bounce Major P's hand came over mine on the eight throttles and he pushed them forward to maximum thrust. We were about 40 feet over the ground nearing minimum flying speed, slowing toward stalling speed. With power applied, the plane seemed to hang in the air shaking slightly, then we accelerated.

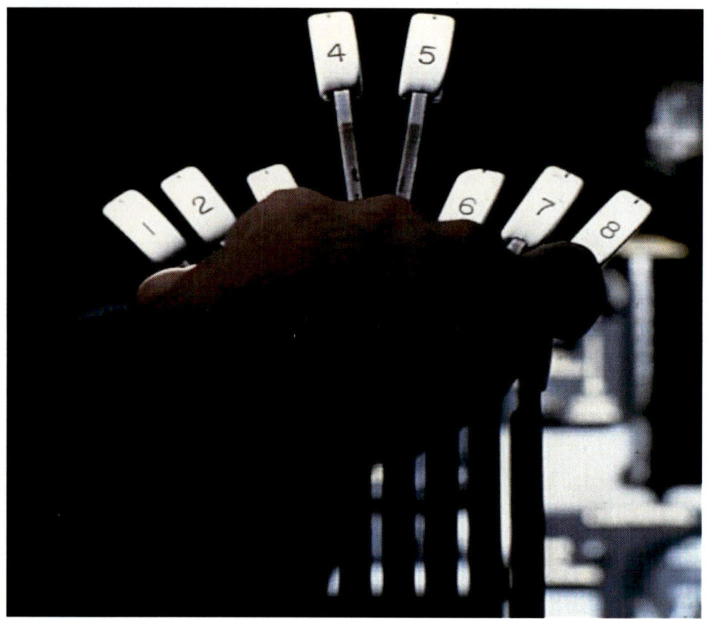

Eight Engines, a Handful

Without Major P's response the plane would have probably stalled and we probably would have had a hard landing, or worse. He kept us alive and I learned from the new experience. I wonder where he is today.

After flight school, I took my white helmet and custom-made mask with me to Beale. I trained and became combat ready. The Vietnam War was ongoing. I flew my first Arc Light (B52D) 140-day tour with the white helmet. We flew out of Guam and Thailand on combat sorties over South Vietnam, Cambodia, and Laos.

Aircraft were being shot down with crewmembers parachuting into jungles. SAC apparently figured out that if you were shot down, a white helmet glowed in the dark jungle. SAC then added camouflage tape to the helmet. My next couple of Arc Light (B52D) & Bullet Shot (B52G) deployments were with camouflaged helmets. Years later the helmets were painted grey.

The Strategic Air Command policy was to keep the six-man crews together for the long term. It was a good policy as each crew became a team. We lived, partied, and flew together (and sometimes died together). We knew each other's strengths and weaknesses. In flight, we coordinated over the aircraft intercom, I knew what the others meant when they said something, by tone, pitch, and words.

Having said that, crewmembers moved and changed. SAC moved personnel from base to base every three years. Upgrade training occurred. Copilots upgraded to be pilots (Aircraft Commanders), Navigators became Radar Navigators. All positions became instructors.

This was a time before computers, the email, iPads, or iPhones. The ground, flight, and alert schedules were developed on the scheduling office wall with a grease pen. Our squadron lives were controlled by this schedule. It depicted our alert coverage, post alert crew rest days, and the ground/flight training schedules.

Our two leave (vacation) periods were also scheduled by crew. Our master schedule included two leave periods a year; one for 20 days and the other for 10 days. So, there wasn't any opportunity to pick your vacation times.

Paper copies of the weekly schedule were made for distribution to each crewmember. A master copy was typed, mimeographed, then the 10 pages were stacked by page number on a table. A scheduler would come into the squadron crew flight planning room and ask for volunteers to help assemble the paper schedules. The volunteers walked around the table picking up each page of the schedule in turn, then stapled and stacked the completed schedule. We then walked around the table again making another copy until all the paper schedules were assembled. Completed, the schedules were then placed in each crewmember squadron mailbox.

The biggest driver of the schedule was a 7-day nuclear alert tour. Alert crews changed over on Thursday mornings. Four days of crew rest followed, Thursday through Sunday. Then there was ground training and/or training flights.

We averaged 3 nuclear war training flights a month. The 10/11-hour training flights occurred on a three-day cycle: Plan, fly, and debrief. To make this three-day schedule work, the navigator would complete the flight form 200. F200 was a detailed flight plan/log with all times, coordinates, altitudes, speeds, and leg lengths. He would then give it to me and I would calculate fuel use for each leg. Gross weight was then used to calculate takeoff data.

On mission planning day each crew position attended specialized study. After the specialized study was complete, the crew assembled and a crew briefing was given. Timing, frequencies, and activity was reviewed.

A typical flight would include an air refueling, a navigation leg, and low-level flight with simulated nuclear weapon releases. Finally, the pilots practiced instrument approaches and landing.

The debrief day completed the cycle. We discussed and critiqued the flight activity covering the good, bad, and ugly. Significant learning occurred during this feedback events.

Each crew spent an average of 10 days on nuclear alert and flew three training flights a month. There were 4 crews on alert at Beale and one crew on satellite alert. Through the years, Beale crews sat satellite alert at Mountain Home and at Hill AFB, UT. Satellite alert spread our nuclear forces around the country which made targeting harder for the Soviet Union.

Four crews lived the same schedule throughout the year. As a result of living 24 hours a day on alert for seven days, we knew the other three crews very well. The other squadron crews less.

Normally, only 4 bombers and crews were on alert, however, the entire wing could be directed to generate to nuclear alert status. A recall of the entire wing could be directed by Headquarters. Within the squadron, there was a published recall roster (plan). During this time frame, we had a phone recall roster of who called who. Then if required, it was followed by a drive door to door plan.

The B52 crews trained to fly and fight a nuclear war. It was called the Single Integrated Operational Plan (SIOP). All SAC units were assigned sorties covering specific targets. My last event to become combat ready was to "Certify the SIOP" by studying and briefing a specific unit assigned sortie. Sortie study lasted many days. Each classified sortie bag contained specific sortie information. Charts and maps of the routing, both high altitude and low level. Air refueling locations and information. There was bomb run information (headings/timing). Finally, there were post-strike charts and recovery information.

Usually a 6-man crew would certify with each member briefing on his specific position. The aircraft commander would start the briefing, then the copilot, navigator, radar navigator, electronic warfare officer, and finally the gunner. A Colonel (Wing Commander, Vice Commander, or Deputy Commander for Operations) received the briefing. There were usually 20-25 wing staff members from the various functional specialties seated behind the Colonel.

When the certifying crew finished briefing, the Colonel started asking questions. Then questions came from the staff. The Q&A went on for quite awhile. At some point, the Colonel was satisfied we knew and understood the sortie. He signed certification certificates for each crewmember. If you failed to certify you returned for more sortie study and then attempted to certify again. We passed and I became combat ready copilot. Finally!

I was assigned to my first combat crew. Gunners back then were senior NCOs. I recall the informal seniority in the plane was aircraft commander, then gunner. Paul was my first gunner. He helped me understand the SAC system and aircraft a bunch! A few years later we were getting, what we called, mini gunners, new to the Air Force and just out of initial B52 training.

SIOP Certified, I started to pull 7-day nuclear alerts. We lived in the alert facility near the nuclear loaded aircraft. The facility was built half way into the ground. Downstairs were bedrooms, bathrooms, showers. Upstairs was a large briefing room, a cafeteria, library, briefing rooms, and crew control. There were ramps leading up toward the alert aircraft from the bedroom level, and ramps leading down from the upper level.

When the alert Klaxon horns blew, crewmembers and ground support staff charged out toward the alert aircraft. The Klaxon signaled us to start engines, copy, and decode an Emergency Action Message. On alert each crew covered specific targets. We needed to be able to takeoff within 15 minutes of notification. Keep in mind that ICBM missile flight time from the USSR was 30 minutes. The idea was to fly the aircraft away from the base before soviet missiles arrived and destroyed it.

I averaged about 10 days a month on nuclear alert at Beale, Mountain Home, Idaho, and Hill AFB, Utah. About 30% of the B52s across the country were on nuclear alert.

The alert facility and aircraft ramp were secure areas. M16 armed guards controlled the entry gate into the alert area that contained the alert facility and aircraft. Also, there were point guards on the ramp at each alert aircraft. There was a red line around each nuclear loaded alert aircraft. Cross the red line and lethal force was authorized. The aircraft point guards were given a name list of the alert aircraft crew and crew chiefs assigned to the aircraft. We gained access to cross the red line in two ways. During normal operations to access the aircraft you gave the guard your alert badge. He checked the access list and cleared you across the red line. During a klaxon alert response, there was a hand signal given between the guard and the responding crew. Security around the aircraft or into the aircraft was controlled by two policies. The purpose of each was to assure that there was no time a single person could have access in or around the aircraft.

The "Two Man Policy" required that at least two men be allowed to cross the red line around the aircraft. Individuals could monitor each other. The crew chief and maintenance could accomplish required work around and under the aircraft.

The "Two Officer Policy" required at least two aircrew officers to climb up into the aircraft. There were Top Secret documents in the plane.

The alert pad was a serious area. Alert B52 aircraft were changed every 45 days. The new B52 was towed into the alert area. Nuclear weapons, flares, chaff, and .50 caliber guns were loaded. Then the alert crew conducted a nuclear weapons and aircraft preflight. The crew then transferred the Top-Secret documents from the old alert aircraft the new aircraft. Aircraft changeover was then complete.

One night we were changing alert aircraft and were making the final move with the Top-Secret documents from the old alert aircraft to the new alert aircraft. The gunner was driving the crew in the Fast Reaction alert truck. We screwed up, we should have had the truck interior lights on so the guards could see into the truck cabin. Our interior lights were off.

As we drove across the parking ramp a guard whistle blew and the gunner stopped the truck. The M16 armed guard ordered us to exit the alert truck on the side he was on with our arms in the air. The six of us stood in a line with

Crew Responding during an Alert

our arms raised. The guard then asked the driver (our gunner) to step forward ten feet and lay his alert line badge on the ground, then step back. The alert line badge contains his picture and his service number.

The guard picked the line badge up and looked at it. It just so happens that the air force was converting from a serial number to our social security numbers. He had just converted. His social security number was on the badge. The guard asked him to state his badge number. The gunner rattled off the old number! Wrong. I could see the guard really tense up. He refocused the M16 on us, the safety came off, and he gripped the rifle tighter.

Still standing in line with our arms up, I hoped none of the others did anything stupid because I think the young guard would have hosed us! This all occurred within a couple minutes. The guard had called a “Helping Hand” over his radio and security backup soon arrived. The situation was resolved and we continued to change the alert aircraft, this time with our interior lights on.

Lightning Strike

When my combat crew flew nuclear training sorties my aircraft commander and I (copilot) took turns making the approach and landing. This night, returning from a training sortie, it was my turn. The weather was terrible with heavy rain, lightening, and with extreme turbulence.

I’m flying an Instrument Landing System (ILS) approach through the weather and the night. There are two transmitters on the ground that provides ILS guidance into the cockpit. One is the Localizer that provided right/left guidance to align the aircraft with the runway center line. The other is a Glide Slope transmitter that provides descent guidance (2.5 degree) down to the runway. Over the ground, that’s about 750 feet per minute descent. In the cockpit, two ILS needles are displayed over the aircraft attitude indicator.

Ever been in a B52 cockpit at night? Flying at night was more challenging than a day flight. The cockpit was kept dark. Overhead lights were turned off. The panel back lights and flight instruments were turned down to the lowest level possible. Checklists were hard to read. It was difficult to see the data on low level maps. The reason for the low light was to increase our night vision so we could see outside the cockpit. It takes about 30 minutes to achieve maximum night vision. I used a flashlight with a red lens if needed in the cockpit.

Strapped into a metal ejection seat, my seat was my survival kit. The parachute was on my back. I wore a flight suit and gloves that were made of Nomex (fire resistant material). My oxygen mask was connected firmly over my face to my helmet. A clear plastic visor was lowered over my eyes onto my mask.

My left hand controls the eight throttles. My right hand on the control column. During the instrument approach, my main focus is to the attitude indicator and airspeed indicator. The attitude indicator displays a gyro created horizon. This instrument is important because our human land-based sensors do not work flying. On the Attitude Indicator, I adjusted the aircraft attitude, up/down or turning right/left, to keep both ILS needles centered.

I was on the runway centerline and on the glide path. The ride was extremely rough. About 5 miles from the runway (1,200 feet over the ground and descending 750 feet per minute), static started to be heard on the radio. Ever heard static on your AM radio during a storm? The static quickly increased to an extremely high level, then there was a brilliant flash with a loud bang.

We took a lightning strike! With the bright flash, I was totally flash blinded. I'm flying the approach and couldn't see anything, I mean nothing! I opened my eyes as wide as I could as if that might help but I still couldn't see anything. After what seemed forever (probably seconds), I began to see circular shapes of the instruments, then finally the flight instruments.

I said, "I have the gauges, and have no flags." (A red flag on a flight instrument meant an inoperative instrument.) The aircraft commander response, "I don't have any flags either so all the flight instruments are good." I continued flying the ILS to a landing. We unloaded the plane in the driving rain and rode the crew bus to the squadron.

We were in the squadron flight planning room filling out the post flight paper work. A maintenance member entered and asked, "Who landed the last plane?" We acknowledged. He said, "Your plane is dumping fuel onto the ramp!" I responded, "We took a lightning strike on final approach."

My take: There are 12 fuel tanks on a B52. The lightning strike created a leak in a tank that was full of JP-4 (jet fuel). If it had been in an empty tank, the fumes would probably have ignited, and the plane would have blown up. Also, the driving rain probably helped to prevent a fire.

The B52 Tail

We finished flight planning for the flight the next day. We left the building and were walking along the sidewalk. Across the street was the B52 parking ramp. B52 vertical tails were proudly sticking up into the air. One was different. The vertical section of the tail was laying over on top of the right horizontal part of the tail. This was the first time I had seen this. My navigator commented, “How do they do that?” That was my cue to screw with him. I said, “There is a big zipper inside the tail. It’s like your pants zipper. But huge. It takes four guys to pull it open or closed.” He bought my BS!

Castle Fog

My crew received deployment orders to the war in South East Asia. At Beale, we flew the B52G model and trained nuclear war tactics. I was a copilot. This was my first crew and first deployment to the war. We would be flying a B52D model with operation’s code named “Arc Light.” We travelled to Castle AFB, Merced, CA, to attend B52D model transition training.

Our training consisted of B52D systems academics followed by three training flights. Three training flights were flown practicing conventional bombing tactics with 3 B52D in formation.

On the third, and final training sortie, the three-ship formation flew directly over Castle AFB toward the south where we would separate, turn north, and begin our individual approaches and landings.

The weather broadcast reported significant ground fog and low visibility. Looking directly down we could see the base and runway through the ground fog. Flying down final the slant angle visibility (forward and down) would be different. The ground fog topped at about 100 feet above the ground. Above was totally clear, below visibility was zero.

As we flew south, the three B52 separated so each could, in turn, make an Instrument Landing System approach to land. The first aircraft descended to the ILS Decision Height (200 feet above the ground) and looking forward couldn’t see the runway.

They made a missed approach climbing above the runway. They were in clear air throughout the approach. The top of the fog was still 100 feet below them but they couldn’t see the runway because slant vision made the visibility worse than looking directly down.

A couple minutes later we were the second aircraft

on final approach. I saw the first aircraft make the missed approach. The control tower could be seen sticking up through the solid white sea of fog.

Wing tip vortices (tornadoes caused by high pressure under the wing interacting with lower pressure above the wing) are generated as the wing tips sliced through the air. These two tornados like vortices then descend. The wing tip vortices of the first aircraft descended into the ground fog over the runway stirring the ground fog up.

We saw the runway when we arrived at the 200-foot Decision Height and landed. As we rolled out along the runway the dense ground fog closed in around us and we had to stop on the runway.

The third B52 was a couple of minutes behind us. They had to make a missed approach over us because we were stuck in the fog on the runway.

The control tower cab was above the top of the fog. They could see our aircraft tail sticking up through the top of the fog. We sat there on the runway unable to see to move until a truck arrived in front of us and slowly led us off the runway to parking.

The other two B52 diverted south to March AFB near Los Angeles and spent the night there.

My crew completed B52G to D model transition training in three flights. Additionally, we practiced three aircraft formation flying conventional tactics used to drop iron bombs.

Training complete, my crew met at the Officers Club for dinner. When dinner was over, I returned to my room. I was now alone with my thoughts. In the morning we would ride a KC135 tanker to Guam, then continue onto Utapao, Thailand. I would be dropping bombs on people within the week. Given the opportunity, the enemy would be trying to shoot me down, capture, torture, or kill me.

I thought back to when I first arrived at Beale, about a year earlier. Attending my first Squadron Commanders Call at the Officers Club, walking past rows of combat aircrews, I suddenly realized I had actually joined a combat squadron. Now it was 4 December 1971, I had been watching and reading about the Vietnam War for years. I suddenly came to understand that I would be going into combat there.

I realized that I would soon be in the kill or be killed insane/irrational war world. I was a bit apprehensive. I decided to make a final rational act before deploying to the unknown. I called and talked with my sister (Bev).

I would ride a tanker that would take me to the combat theater in a few hours. I wondered what my Dad or my Uncles were thinking about as they deployed during WWII. I travelled to the combat zone in a day. They rode packed troop ships on 7 to10 day journeys. The ships sailed in convoys where the speed was dictated by the slowest ship. They lived with the threat from enemy submarines.

Dad once told me a WWII experience. (Appendix 1) He sailed on a Dutch ship. They were fed boiled kidney. Troops were packed in, sleeping in hammocks stacked four high. During heavy seas, many of them became sick. He said vomit sometimes sloshed back and forth across the compartment floor as the ship pitched and rolled. Those were different times. The next morning, we rode a KC135 to Guam.

Close Encounters

My crew deployed Arc Light (B52D) to South East Asia (SEA) from 5 Dec 71 to 28 Apr 72, 140 days. My first combat tour in (SEA) was flying sorties out of Utapao, Thailand. A number of exciting events occurred as we flew 66 combat sorties.

Our mission for the night was a target in the Plain of Jars, Laos. We had been targeting the same area for the last week. Night after night we continued striking the same general target area, a place called "The Skyline Ridge." The CIA backed forces and the Hmong were charged to hold the ridge. Time after time the communist Pathet Lao forces would take it. We would bomb along the ridge and CIA forces would retake it. Then Pathet Lao forces would overrun it again and again and take control once more. This battle went on night after night.

We were dropping Cluster Bombs Units (CBU's), they are an anti-personnel weapon. The bombs fell to a pre-determined altitude, then like a clamshell split open releasing hundreds of baseball size munitions. The munitions detonated as they hit the ground sending ball bearing like shrapnel in all directions. This was definitely a place that you wouldn't want to be around.

We carried 44 CBU's internally in the bomb bay and 24 externally under the wings. The aircraft was also loaded to defend itself. We carried 96 flares to defend against heat seeking missile attacks. There was radar defeating chaff that could be dropped. Finally, we carried 2400 rounds of ammunition for our four 50 caliber tail guns. We felt that would be enough fire power should we encounter any aerial enemy resistance.

B52s flew in three aircraft formations we called a cell. It was an all-out effort to launch three aircraft. Hundreds of personnel were involved in this massive undertaking. Intelligence teams prepared the strike folder information and briefed us on the mission. Maintenance, munitions, and fuel personnel prepared the aircraft. Aircraft crew chiefs oversaw the maintenance preparation of their aircraft. They then assisted the crews in the preflight and launch of the aircraft.

B52D releasing on a target

To launch a three B52 cell, five aircraft were prepared with full combat loads which consisted of three primary aircraft and two spare B52s for backup. While the three primary flight crews completed the numerous sortie briefings and target study, a fourth crew, the preflight crew, was responsible for preparing the two-spare aircraft. They would start one spare aircraft as we started the three primary strike aircraft to become the engine running spare. The fifth preflight aircraft remained an unmanned aircraft, cocked and ready for an immediate start if needed.

The pre-sortie crew briefings were many. The three six-man crews initially received individual briefings by their specialty areas.

Pilots studied the timing, routing, fighter support,

frequencies, the target heading and timing, and the post target turn. Radar navigator and navigator accomplished the same basic briefings with a focus on the route, radar offsets, and target study. The electronic warfare officer and gunner studied ground and airborne threats.

Everybody had their own areas of expertise and was expected to know their individual responsible areas. We flew as a crew and if things went bad could die as a crew. Our lives could literally depend on the actions of fellow crewmembers.

After the specialty briefings, the three crews assembled for a mass strike briefing. Finally, after the mass briefing was accomplished, the cell leader reviewed the formation, airspeeds, action points, and tactical cell procedures.

Every crew action was based on a predetermined timing schedule. There was time scheduled for a crew bus pickup, briefing, and arrival at aircraft, start engines, a takeoff, and a weapons release time.

Everything in this mission was dependent on precision actions and precise timing meaning that all crews were dependent on each other. We arrived at our aircraft one and a half hours prior to the scheduled takeoff time. Each crew member had specific preflight actions to accomplish.

I preflight my ejection seat and parachute. Then according to the checklist, I accomplished cockpit systems checks, and inspected my assigned parts of the aircraft exterior. Each member of the crew conducted similar inspections of their assigned areas. The navigator team pulled the safety pins from all the bombs.

This night we were scheduled to fly the number two position in the three-ship cell. The cell leader was always an experienced crew. The logic of this doesn't need an explanation. The number three crew this night was FNG's (Friggen new guys). This was to be their first actual combat mission. It was safer to have new crews in the number three position as they gained operational experience. (You will understand why later!)

Four aircraft started engines according to the timing plan, three primary strike aircraft and the engines running spare. The primary three aircraft taxied to the runway hold line. The lead took off exactly on time. One minute after lead departed, we took the runway for takeoff. The aircraft commander pushed the eight throttles up to set takeoff power. The number seven engine fire light immediately illuminated! This glowing red light was an indication of a possible engine fire!

The throttles were reduced to idle and the light warning extinguished. We had to shut down the engine immediately. The light could have been an actual fire or a malfunction in the warning system. Either way, we couldn't fly this aircraft without the issue being looked at. We aborted the takeoff and turned off onto the parallel taxiway where we could complete the emergency engine shutdown checklist. The number three aircraft (the FNG's) took off one minute after we aborted.

With our abort radio call, the ramp turned into furious and frantic activity. My crew needed to move to the engines running spare, takeoff, and catch the other two aircraft. The engines running spare crew taxied the spare to the runway hold line as soon as we made our abort call. We taxied our broken plane up behind them. Our crew proceeded with an engine running "bag drag" (rapidly moving to another aircraft with all our bags) to the engines running spare.

I held the aircraft brakes from my right seat as my crew frantically packed up their personal gear, strike materials and made the move to the engines running spare. I could see the white helmets of my crew dart out from under the aircraft nose and run forward to the engine running spare. A nameless pilot from the engines running spare climbed into the pilots left seat and took control of the plane from me.

It was a dark night of frantic activity. I left and moved to the new plane. This all occurred in the dark of the night and in mass commotion. Fifteen running engines assured us that no one on the ramp could talk.

We grabbed up our personnel equipment, classified strike maps and information, and then we were up into the new plane. Strapping into the ejection seats with shoulder straps and a lap seat belt all done by feel in the darkness. All the time the pilot pushed us all over the intercom by asking repeatedly if we were ready for takeoff.

The bag drag to the new aircraft occurred in just a few minutes and we took off into the night. The other two aircraft were just minutes ahead. The only way to catch them was with increased speed and the use of the timing triangle. We finally caught the other two aircraft just as the bomb run began. We slid into number three cell position one mile in trail and five hundred feet above the number two (the FNGs).

The cell formation structure was each aircraft one mile in trail stacked up 500 feet. Number two was 500 feet to the right and 500 feet above lead and we were 1,000 feet above and 500 feet to the left of lead. Over the ground, we crossed any point at a six second interval. The goal was to lay three parallel 500-foot bomb trains onto the target area.

Our target was in a known MIG threat area. MIG21s had made missile runs on other B52s in this area on previous nights so we had to always be on our guard. We had MIG Cap assigned to protect us. Air Force F4s rendezvoused with us at the bomb run Initial Point. They were out in the darkness somewhere paralleling us to come to our aid if we needed assistance. It was a simple radio call to them and they would be by our side in a matter of seconds. Their call sign was "Gunfighter," ours "Red Cell." "Gunfighter, Red Cell plus 5" .... We were 5 seconds late. They responded with a "Roger Red Cell." It was comforting to hear they were out there watching over us.

The tactic in any threat area is to minimize the time in the threat area. The tactic in a MIG threat area was that at thirty seconds after bombs were released, a 45-degree bank angle right turn was made to a post-strike heading (135 degrees to the right) was executed to depart the threat area. With each aircraft turning at the same point over the ground (six second intervals), all would rollout on the post-strike heading in trail of the cell lead. This turn would get us heading away from the MIG threat area as quickly as possible and back to safety.

We arrived at our number three position at the Initial Point of the bomb run...about 6 minutes to bombs away. There was no time to adjust the cell so we could be in our planned number two position. I maintained our distance from number two (the FNGs) by adjusting power with the throttles. The pilot used minor heading changes to keep our lateral position from the cell lead position.

It was a clear night and during the bomb run, looking down and ahead, I could see fires burning across the ground in the target area. There were artillery flashes and tracer streams across the entire area. The sky lit up like the fourth of July. It looked really brutal and nasty down there on the ground and I recall thinking how I was glad that I wasn't down there.

Ahead a number of fighters were already busy engaging targets in the same area. There were fighter aircraft forward and high to the right of our target area. Running lights could be seen as the fighter flight organized.

Then the lights went out as the fighters dived to make their bombing runs. I couldn't see the fighters but knew where they had just been. Triple A (Anti-Aircraft Artillery) rounds could be seen streaking upward followed by the detonation flashes. Triple A fire swept from the right to left across the dark night sky as the enemy gunners on the ground attempted to shoot down the passing attacking fighters. Post-strike, the fighters' running lights appeared forward to our left as the fighters reassembled for another attack.

Our cell was now on our bomb run and the ops tempo in the plane increased. It was really busy but each crewmember knew what needed to be done and when.

There was nothing else to be done but business on the interphone and radios. The navigator team ran their respective bomb run checklists assuring that all switches were in the correct position to affect a release. The navigator called out each checklist item; the radar navigator confirmed the switch position and responded verbally with the position.

The radar navigator identified and took the final bomb release radar offset. Each aircraft would cross the bomb release line in trail at 6 second intervals. The radar navigator could see the other two aircraft as blips on his radar scope and when able gave us an abbreviated narrative of what was taking place.

"Number one is releasing," "Number two should be releasing." The pilot's time-to-go meter counted down the seconds to our bomb release. Then over the interphone came the radar navigator verbal countdown timing, "20, 15, 10 seconds, bomb doors (bomb door open light came on), 9, ...3, 2, 1, Bombs away."

A "Bombs Away" light at the pilots' station rapidly flickered as an electrical impulse released each bomb in turn. The aircraft shuttered/trembled slightly and the aircraft trim wheel spun rapidly in the nose down direction as the enormous weight of the bombs rippled off the aircraft.

Shortly after our weapons were released the radar navigator watched the lead aircraft on his radar begin its post target turn. He called over the intercom that the lead aircraft at 30 seconds after its release, beginning its 45-degree banked turn to the right toward the briefed post target turn heading. He then stated that number two (the FNGs who were planned to be in the number three position) should have started its turn at their 30 second point, but they didn't! Rather, number two continued straight! This was a big mistake and a big problem. Where were they going?

We arrived at our 30 second timing and began our 45-degree banked turn to the right. The autopilot kicked off in the 45-degree banked turn. Our plane lost and then gained a couple of hundred feet altitude. While we were in the turn, the radar navigator announced that the number two aircraft had now started their right turn.

Now they were coming at us from our left, somewhere in the darkness. It might mean that soon two aircraft could be turning through the same airspace as each turned toward the post target heading.

Within my plane it became very quiet…no one talked. Each crew member was alone with their own thoughts. Suddenly it was becoming very lonely. I'm strapped to an aircraft in a metal ejection seat that had an armed explosive charge intended to launch the seat upward.

Bombs Impacting

My seat is a survival kit…that's what the seat contains. On my back is a parachute. There were straps that connect my survival kit (my seat) to the parachute on my back (the back of the seat). I'm held to the metal ejection seat with a seat belt and shoulder straps. I have a helmet on with the oxygen mask connected tightly to my face. My contact with everyone in, and outside, the aircraft is via the interphone through the headset in the helmet and mic in the mask.

It's become totally quiet in the plane. We all know that this is very, very, bad and that almost nothing good could come of all this. An aircraft is coming toward us through the darkness, and that at any second, we could hit and the planes, and all that are in them could be plummeting to the ground.

I looked down to the right in our 45-degree banked turn to the ground below. All I see are the fires burning on the ground, the artillery flashes, and the ground tracers. I think to myself, there will be a midair collision and if I survive it will be my own guys (the FNGs) who will be putting me into the combat hell below.

I prepared to eject from the aircraft assuming the ejection position. My helmeted head back against the head rest, feet back against the seat on the floor, elbows tucked in, and hands on the ejection triggers. I looked down to the right into the night at the hell below, waiting in silence, for the expected midair collision.

After what seemed like forever the number two aircraft (the FNGs) reappeared on the radar to our right! We missed! Somewhere, somehow, the two aircraft passed each other without a midair collision.

Above, below, in front or behind, I don't know. But the two aircraft missed each other! God was with us this night. A total catastrophe was avoided and we still don't know how it happened, but obviously we were very glad that it did.

The cell reformed behind the lead aircraft. The flight back and landing at Utapao Air Base was uneventful and routine. After landing the three crews met and did an extensive mission debriefing. The FNG crew knew they had screwed up the turn timing and had put all our lives in jeopardy. Luckily the night turned out ok and we all survived, the FNG crew survived, and we all learned a valuable lesson from all this. After that, we flew a number of other combat sorties with them (no longer FNGs) and they did just fine.

And there you have it. We survived another night combat sortie over Laos. But wait, this story doesn't end there.

Remember that on our first takeoff attempt we aborted because of a number 7 engine fire warning light. We bag dragged to the engines running spare. We moved from an aircraft that we had prepared, preflighted and moved to an aircraft that had been preflighted and prepared by the duty crew.

As we were shutting down the eight engines, I reached down to unbuckle my ejection seatbelt and shoulder straps to release me from the metal ejection seat. In the early light of the morning, I noted that the straps connecting my survival kit seat to the parachute were incorrectly threaded through my seatbelt, effectively connecting me permanently to the ejection seat. It was a sobering find! The preflight duty crew copilot really let me down. If I had ejected from the aircraft during the post target turn over Laos, I would not have separated from the ejection seat and there would be no parachute opening. I would have fallen to the ground from 31,000 feet still connected to the seat……and perished.

Life is always interesting. One can't predict the future. Who wins, who loses. How it is that one person dies and another survives? Why them and not me? Was it divine intervention, superior talents and preparation, or just plain luck? Maybe a bit of all three.

Bonus Deal

Occasionally, a B52 lost its Bombing and Navigational capability. The lead gunner would provide directional guidance to the following aircraft. The term "Bonus Deal" was used when a gunner directed the following aircraft to the bomb release position.

In 1971, I was a copilot deployed in Arc Light in the B52D model to Utapao, Thailand. The combat sorties were about 2 hours long so things went fast. After level off, each aircraft Bomb Navigation System (BNS) was checked. On this night only one of the three aircraft had a functioning BNS bombing capability.

The capable B52D moved to the lead position. The lead gunner then used his aft facing gun radar to accomplish a double Bonus Deal by verbally guiding the two-trailing aircraft through the bomb run and to release.

He was awesome! I had seen a number of single Bonus Deals before but this was my only double! The lead radar navigator called out the wind drift. The lead gunner called out range (2,000/4000 yards) and azimuth (number degrees right or left) to the two-trailing aircraft. The following two pilots made minor heading adjustments to be in the correct lateral position while the copilots adjusted power to maintain the correct range. The lead gunner successfully directed the two-following aircraft through weapon release. This intense event was completed calmly and professionally because of training and experience. The leader gunner then led us back to the base.

Piggy Backs

We carried forty-four 750 lb. bombs internally in the bomb bay and twenty-four 500 lb. bombs external under the wings. Occasionally, all the bombs didn't release.

As the weapons rippled off a B52D the aircraft shuddered and the horizontal stabilizer trim wheel, located next to the pilot's right knee, spun rapidly in the nose down direction as the autopilot strained to adjust the trim to keep the aircraft in level flight. The leading edge of the massive tail horizontal stabilizer moved up or down adjusting for aircraft weight and speed.

There was a single "Bombs Away" light on the pilot's cockpit instrument panel. It flickered rapidly, too rapid to count, as each bomb received a release impulse and dropped. The light told me that weapons were releasing.

Downstairs at the Radar Navigator/Navigators station there was a panel of lights, one light for each bomb. Each light indicated the specific location of a bomb, internal and external. Each light went out when its bomb received a release impulse.

On this night sortie the cell of three B52Ds flew the bomb run and released bombs. At bomb release, our cockpit "Bombs Away" light flickered rapidly. During the post target turn the Radar Navigator announced over the intercom that one bomb apparently didn't release! One of the bomb indicating lights stayed ON. It was a bottom weapon in the bomb bay. This was a potential problem. The bomb bay racks were fixed racks that were spaced with just enough room for each bomb on that rack to release, slide down between the racks, and drop from the aircraft. If the bottom weapon didn't release and the three above did then potentially, we had three 750 lb. bombs stacked up on top of the bottom weapon! These were called "piggy backs."

The loaded bombs were not armed. Each bomb was double fused, one nose fuse and one fuse near the tail. The fuses had a small propeller that spun as the bomb fell through the air. After a given number of revolutions, the bomb became armed. Each bomb fuse was safety wired to the aircraft. Stiff wires attached to each fuse preventing the prop from spinning and from being armed until released. When the bomb released the safety wires pulled out of the fuses. Piggy back bombs released and dropped so their safety wires were pulled. With a correct air flow, the bombs potentially could arm. Not a good thing.

Sometimes the bomb light indicators would malfunction so we needed to confirm our situation. The

Three Clips fit into the Bomb Bay, Plus 24 under the Wings

procedure was to level off at 10,000 feet and have the navigator crawl back and look into the bomb bay.

Level at 10,000 feet the plane was depressurized and the navigator opened the small hatch aft in the lower compartment. He then went off the intercom and crawled back through the alternator deck, squirmed through the forward landing gear to another small hatch at the forward end of the bomb bay. He was off the intercom and out of sight for most of the journey, not a comfortable event. It was night, he crawled with a flashlight through the tight squeezes.

The radar navigator was at the compartment hatch watching with a flashlight giving us a commentary on the navigator's progress. It seemed like the navigator was out of sight a long time.

The navigator returned to the pressurized compartment and came up on the intercom. He confirmed, "We have one hangar and three piggybacks!" We informed the command post of the situation and declared an emergency for landing.

The other two nonemergency B52s landed first and cleared the runway. It was now our turn to land. The moment we touched down the radar navigator reports, "The hung bomb light just went out!" This indicated that it released and we now had four 750 lb. bombs rolling around on top of the bomb doors!

I quietly hoped none had broken through the doors and dropped onto the runway. Having a 750 lb. bomb tumbling along the runway with you is not a good thing! We stopped at the runway end and were directed to taxi into an isolated area on the hammerhead well away from all the other aircraft.

As we sat there a munitions man crawled up through a forward wheel well and made his way to the small bomb bay hatch. He confirmed four weapons were laying on the doors. We shut the aircraft down, gave the plane to maintenance, and got the hell out of there!

Dropped the Drag Chute

Drag Chute helps to stop the Plane

One night I totally screwed up. I was a copilot on a B52D returning to Utapao, Thailand, from a combat sortie in SEA. We landed late at night. After landing I was supposed to deploy the drag chute by pulling back on the drag chute lever. The chute deployed out of the bottom near the tail. It helped to slow the plane and save the aircraft brakes. Then after turning off the runway the chute was jettisoned by moving the lever to the forward position.

During this landing, I don't know why, but I incorrectly pushed the lever forward (jettison position) then thinking, that's not right, pulled it aft (deploy). The 200-pound bundled chute dropped out of the plane and slid down the runway behind us! We stopped safely but I bought the crew beer that night. Have you ever screwed up?

B52 Rat Status

Sometimes excitement can occur before the flight. The Vietnam War was ongoing through my high school and college years. The war was intense and permeated all the news and society. I recall reading about Tunnel Rats. A Tunnel Rat was a GI that would crawl down into Viet Cong tunnel networks. They didn't know what they would find: booby traps, snakes, or shooting VC. They went armed with a flashlight and pistol. I recall wondering what they were thinking as they crawled into a tunnel head first. They were a special breed of men.

I believe I learned what they were thinking. I don't claim Tunnel Rat status, but I do claim B52 Rat status. For weeks we flew combat sorties 6 days in a row, then had one day off. Each combat sortie was about 2 1/2 hours long. The sorties were pressure packed as there were many inflight checks and actions to complete before getting to the target.

Each sortie was scheduled with a published timing for each event. There was a bus time, a briefing time (specialty briefs, Pilots, Navigators, Electronic Warfare Officer, Gunners), then a mass cell briefing, finally the cell leader briefed. There was then timing for arrival at aircraft, takeoff, and finally time over target time (bombs away).

We were assigned to fly an early morning sortie on 10 January, 1972. At 2:30 am, the strike briefings complete, we left the Operation Center and walked toward our three crew buses. An operations staff member ran out the door behind us and yelled, "Don't leave yet. Something is happening on the aircraft ramp!" A couple minutes later he come out again, waves us on and yells, without any explanation, "Go ahead." Our Thai driver drove us on the perimeter road around the end of the runway toward the B52 parking ramp and arrived at the guard bunker at the ramp entrance. It's on a small hill overlooking the B52 parking ramp. I can see B52D tails sticking up out above of the parking revetments.

An arm and hand extended out of the guard bunker wall halting us. A voice says, "Hold your position, there are sappers down on the aircraft parking ramp!" (Sappers have explosives and intend to attack and blow things up). We see flares popping in the air, and hear automatic weapons firing, Tat-tat-tat, Tat-tat-tat. I ask the bus driver to please turn off all the bus lights, interior and exterior. A tracked armored vehicle goes screaming around us and disappears down onto the ramp. There are more flares and automatic weapons fire. Then it gets quiet. After a few minutes the arm comes out of

B52D Tails above the Parking Revetments

the guard bunker waving us forward, the voice says, "**Go ahead, we think we got them all**."

The bus parked in front of the B52D we were to fly that night. The crew chief climbs in the bus with the aircraft paperwork for the Aircraft Commander to review with the crew. The crew chief told us that sappers were among the B52s and were trying to blow them up. He said they only blew up one engine and then he said, "**Security thinks they got them all**."

With that, my pilot turns to me and says, "Co, go up into the crew compartment and check it!" What! I don't recall if I loaded my .38 pistol. During the 30-foot walk to the plane crew entrance hatch I'm thinking, "**They think they got them all.**"

There are two decks in the B52 crew compartment. The only way up is head first, twice! As I approach the entry hatch I'm thinking, "If there is a sapper up in the crew compartment I'm dead." Ever do something expecting to die? I know what the Tunnel Rats were thinking. I claim B52 Rat status.

Post script: "We think we got them all." Later we learned that there were two sappers. During the firefight one was killed, **but one escaped**. They had snuck through the Air Base perimeter wire intending to blow up B52s with satchel charges. Explosives were thrown into a number 7 engine and on the ground under two other B52s. The three explosions occurred just before we left the operations building for the buses so we didn't hear them. All three damaged aircraft were repaired within 24 hours. One sapper tried to shoot a maintenance man three times but his gun was either empty or jammed.

Screw the Boss

Combat sorties from Guam were typically 12 hours long. There was a lot of boring deadhead time. We would go to the library and record music to play over the intercom while cruising. Back then the technology was reel-to-reel tape.

One day the navigator and I decide to screw with the pilot. While recording a song the navigator records on the tape, "Pilot come 20 degrees right." After about 30 seconds he records, "Pilot come 20 degrees left."

The next day we takeoff on a combat sortie. Looking forward and down, we are 500 feet above and 2 miles behind the number two aircraft. Looking down past number two, I see the cell lead a thousand feet below us.

The taped music played as we droned along. We are in perfect formation position. Over the intercom (on the tape), the navigator voice tells the pilot to turn right 20 degrees. I look over at the pilot, my smile is hidden by my oxygen mask. He turns the plane 20 right. Thirty seconds later the navigator tape voice directs a left turn 20 degrees.

We are now paralleling the other two planes about two miles to the right. To my amazement, the pilot didn't question the turns. You have to trust the navigator. The pilot never knew what we did.

Initial Buffet

Learning to fly a combat aircraft is a daunting task. It takes untold hours of study, countless simulator training sessions, and dozens of training flights. There are tests and check rides to pass. In the end you are certified combat ready. At this point you are turned loose. What is missing at this point is the learning and surviving of an unusual or new experience. I survived a new experience in Thailand.

This was our first Arc Light tour (B52D models in South East Asia). I was a new copilot on my first crew. The Aircraft Commander was also new to the B52. Previously, he had flown a year in an O-2 Bird Dog as a Forward Air Controller in South East Asia. The O-2 was a very small single engine Cessna-172. The B52D was huge with 8 engines and grossed 488,000 lbs.

We were deployed Arc Light for 140 days. From Utapao, Thailand, we flew 6 days/nights in a row, then had a day off. During this deployment we flew 66 combat sorties. The sorties were only about 2 ½ hours long so we flew with small fuel loads. The aircraft weight was light.

According to deployment schedules, crews and aircraft were moved between Guam, Thailand, and the States. My crew was coming to the end of our 140-day assignment and was scheduled to return to Guam. We were assigned a B52D that we would fly to Guam after that night's target was struck. The other two B52Ds would return to Utapao.

All three crews attended the usual pre-strike route and target study, then Mass and Cell leader briefings. Other than we were flying to Guam post-strike, there was nothing briefed or discussed about the impact of the performance differences because of the different aircraft gross weights. We were scheduled as number three. Though not discussed why, probably because of our heavier weight would give us a longer takeoff roll and slower climb rate. We took off 1 minute after the second aircraft.

Lead and number two leveled off well before us. Because a tail wind would put us early over the target, the cell leader called for a slower than planned cruise speed. Our takeoff and climb were normal, that is, until we reached the level off at 33,000 feet. We slowed to the speed announced by the cell leader. As the pilot leveled off, I pulled the 8 throttles back to set the usual cruise power setting we had used for the last 4 months. The plane immediately began extreme heavy buffeting! Think stall and not flying. Without comment, I immediately applied full engine power as the pilot instinctively pushed the nose down putting us into a descent. We needed the aircraft airspeed to increase above the stall speed!

We descended a couple thousand feet and the airspeed increased slightly. The pilot tried to level off by lifting the nose. With this slight G-load increase, we immediately began heavy buffeting again! He pushes the nose even further down giving us an increased descent rate. It's called trading altitude for airspeed, again hoping to increase the airspeed. This time it worked.

With increased airspeed we leveled off, then climbed. We had lost a lot of altitude! Cell leader called for an increase in speed for the bomb run. We caught the other two aircraft during the bomb run and released our weapons as planned. Post-strike, we separated and flew to Anderson AFB, Guam. The other two returned to Utapao.

Analysis: We, the pilot and I, had never flown in a mixed weight formation before. Apparently, neither had either of the other crews since the subject didn't come up. We were heavy, they were light. There was no discussion about weight differences, nor the resulting performance differences, during Mass or the Cell Leader briefing.
Our heavier aircraft would stall at a higher airspeed than the other two lighter aircraft. In other words, they could fly slower than we could. We were at or near buffet/stall speed, but our set climb power kept us flying. At level off when I pulled power the heavy buffeting started.

Had the pilot not lowered the nose into a descent, we probably would have totally stalled. The buffeting was so great I wondered why bombs didn't shake off the plane. While flight training in small aircraft we actually put the plane into total stalls, then recovered. During B52 training, we practiced "initial buffet" where we carefully slowed plane in level flight until a gentle buffeting just started, then recover with full power. I survived a new experience, survived, and learned! I became wiser.

Taxi Crew

On Guam, my crew had a few days off. One day I was sitting on the billeting stairs enjoying the day watching a B52 on final approach to land at Anderson AFB. Something wasn't right, then I realized it had a short tail! B52Ds had a tall tail. The short tail B52 on final was a G-model. It was the first of many. "Bullet Shot" operations deployed B52Gs to Guam to augment the Ds.

Soon all the aircraft parking locations on the Anderson parking ramps were filled with "Arc Light" B52D, "Bullet Shot" B52G, and transit aircraft. B52s needed to be moved around the base for maintenance purposes and the concept of the "Taxi Crew" was created.

My crew was the first "Taxi Crew" on Guam. Our first shift was a 12 hour all night shift. We started engines and moved B52Ds around the ramp from one parking location to another. As the pilot and I taxied them around, our electronic warfare officer followed with a crew truck. After we shut it down, he would drive us to the next aircraft.

During that first night there was considerable confusion in the maintenance community. There are many types of required maintenance that a B52 needs to undergo. All the various specialties wanted the plane to accomplish their individual specialty needs. A single point had not been setup to coordinate all the moves. As a result, we were receiving conflicting guidance about which aircraft to move next. We received direction over the aircraft radios, from maintenance vans, and from the crew chief over the interphone. On one occasion we were directed to a particular parking spot and aircraft. When we arrived the folks in a maintenance van told us to change to a different plane. It was parked next to the one we were originally headed for. We did our quick preflight and started engines in preparation to taxi to a new location. A full Colonel suddenly appeared at the instructor pilot seat and directed us to shut down and move over to the B52D on the next parking ramp and move it! I think the maintenance heavy hitters were squabbling over who was in charge. It was growing pains that occur when setting up a new procedure.

We moved planes around all night. Ground and Tower Controllers controlled access to the taxiways and the runways. We contacted Ground for permission to start, then gave them the taxiway route we planned to take to the new parking location. When we needed to cross or taxi down runway we contacted the Control Tower for approval.

I called Ground with our planned taxi route for his approval. It was fairly lengthy with 8-10 involved taxiways. He approved as requested. Then we started to taxi and three B52s landed and began to taxi into park. Our cleared route was blocked so I called in a revised route, another 8-10 different taxiways. He approved our request. This new route became blocked and I called in yet another extensive change. I could tell by his voice that the Ground Controller was pretty frustrated when he said, "Taxi wherever you want, just don't call me again." And we didn't!

Parking Ramps were totally filled with Aircraft

We moved planes all night. We were taxiing as the sun was coming up and the end of our 12-hour shift was approaching. As we taxied the Ground Controller asked us who we were and to hold our position. We stopped. I figured the ground controllers must have changed over (shift change). I explained, "We are the taxi crew, have been moving planes around all night, and was told by Ground Control that we didn't need to coordinate with them." He responded again, "Hold your position." The radios were quiet for about 5 minutes. We wondered if he thought we were stealing a plane. Would the security police show up? My guess was that he was making phone calls to confirm who we were. I was really tired and told the pilot, "Let's just shut down here and leave." Thankfully, he turned my suggestion down. The controller finally cleared us to move. Our Taxi Crew tour ended a short time later.

At the end of our tour in Guam we returned to Beale AFB on a KC135 tanker. The plane was packed. During our days off, I was able to travel to Okinawa and shop. Things were cheap in Okinawa. I bought a TV, a reel-to-reel tape recorder, plus other stuff. In all, I brought back 14 pieces of baggage. After landing at Beale, the Customs official asked everyone to layout their baggage out on the ramp. I put my 14 bags in a line and waited.

The Customs Agent walked along my stuff. He pointed to a box that had TV picture on it and asked, "Is that a TV?" I answer, "Yes." "Is that a reel-to-reel recorder?" "Yes." And so on, as he walked past my stuff without stopping. The guy next to me had two bags. The Customs Agent digs through his stuff. He pulled out a tube of tooth paste and pushed a wire into it! Go figure.

My first 140-day combat tour ended. I flew 66 combat sorties, learned more than I ever thought I would, and survived some really exciting times! Not bad for a young copilot still new to this military flying life. Back at Beale I was moved to another crew.

Geese

At Beale, we were on the aircraft parking ramp getting the B52 ready for a training flight. There was a solid stratus cloud overcast at about 1,000 feet above the ground. I could hear geese squawking above, I mean lots of geese. I watched a B52G takeoff and disappear into the overcast. A few minutes later the B52G calls back with a bird strike and had lost an engine. Its training mission was over. It was too heavy to land, so it went into hours of holding to burn off fuel to get down to landing weight.

I was still hearing lots of geese above when a KC135 took off and disappeared into the clouds. A few minutes later the KC135 reports multiple bird strikes. Flying was cancelled for the day.

The cloud layer was about 1,000 feet thick. Geese were flying south just above the cloud deck. As the aircraft climbed through the top they flew through wall to wall geese.

Engine Condition Monitoring Program

In the early 1970's, B52 engines shelled all too frequently. Engine compressor blade tolerances were extremely tight. If one blade broke it would cause a ripple effect that results in all the blades to fail destroying the engine. I recall watching a B52G make a takeoff from the ramp. During flap retraction, I saw an explosive flash out on the right wing. An engine shelled, destroyed.

The crew could not continue on the training sortie. They flew around in a holding area for hours burning off fuel until the gross weigh was reduced down to the permissible landing weight. Expensive training was lost and an expensive engine change and rebuild would be needed.

In 1973, an Engine Condition Monitor Program was developed. We were directed to accomplish one on every flight. Using airspeed, altitude, and outside air temperature an Engine Pressure Ratio (EPR) was calculated. EPR is a measurement of engine power (think thrust). The inboard four engines were pushed up and set to the computed EPR. The outboard engines were reduced as required to maintain the target air speed. After a couple minutes all the gages stabilized and the engine data was recorded. Then the outboard engines were set up to the computed EPR, and the inboards were reduced to maintain the desired speed.

The data sheet was then turned in. The data was loaded into a computer program. Trend changes in the gage readings gave a heads up of potential failure. Engine problems could be fixed before engines totally failed. The problem of shelled engines was reduced to almost none.

I was selected for promotion to Captain in August 1972.

I was an experienced copilot and was assigned in the Standardization/Evaluation Division. I would be scheduled to fly with new pilots on their first B52 solo flight. They wanted a seasoned, experienced copilot with the new pilot.

On one flight the landing gear didn't retract when the gear handle was moved up. We were scheduled to accomplish an air refueling, a low-level route, and then make a number of low approaches, followed by a full stop landing.

We climbed to the designated altitude with the landing gear hanging. After level off, we called down to

Beale on UHF radio and discussed our gear issue with maintenance. They suggested we try a few procedures to raise the landing gear. I said to the new pilot, "There is obviously a landing gear malfunction. What happens if you manage to raise the gear, but they don't come down later? I suggest we keep them down, land, and let the maintenance work the landing gear problem on the ground." That's what we did.

Linebacker II

In 1972, I was a copilot assigned to the Standardization/Evaluation Division (Stan/Eval). Dick was my pilot. At that time, I had about 900 hours experience in the B52. We were evaluators who tested and gave check rides to the Squadron crewmembers. We were the black hat guys!

Being a copilot, I checked paperwork that the crew calculated: weight and balance data, the Form 200 fuel calculations, and takeoff/landing data computations. I couldn't give any check rides in the B52G but I could administer pilot/copilot checks in the simulator. In doing that I made many trips down the valley to give simulator checks at Castle AFB, Merced.

Linebacker II (The 11-day war) air campaign over North Vietnam occurred in December 1972. B52s were being shot down. There were only three of twenty crews remaining at Beale. All the others had deployed to Guam and Utapao. There was only 1 bomber on nuclear alert, normally there were 4. One of the three crews manned the alert bird. We also accomplished B52F to B52G transition training to crewmembers coming out of the Castle B52F who were moving to a G model base.

Earlier in the day my parents called. We talked awhile then mom asked, "Do you think you're going to deploy to Guam?" No doubt they saw on the news that B52s were getting shot down. "No way, there are only three crews here, I'm needed to sit alert, and transition crewmembers into the B52G." I clearly heard relief in her voice.

Oops! Because later that evening, when I returned to my apartment at 10 pm there was a note taped on my door. It said, "Fly B52 to Guam tomorrow. Show time 2 pm. Pack for the duration." (Technology was different back then. There were no iPhones, iPads, calculators, computers, or emails. I had a telephone in my apartment.)

Wow! I ran down to the apartment manager and told her I would be leaving for an unknown length of time. I then went into a fast pack mode. Not knowing what I would be involved in, I packed equipment needed to fly both conventional weapons and also needed for nuclear alert.

Two B52G aircraft were being flown to Guam to replace ones that had been shot down. A Mather crew flew the other plane. We took off at 5:30 pm on 21 December. After flying 14 hours we landed at 1:30 am on the 23rd. We were driven directly to a hotel off base and told not to leave until it was determined what they were going to do with us.

Unknown to me there were two issues being sorted out. The local schedulers had us plugged into the Linebacker II combat strike schedule. On the other hand, the

800 Foot Cliff off the End of the Guam Runways

state side folks needed us to return to Beale and resume the F to G transition training to keep the aircrew pipeline flowing. I didn't know any of this until late afternoon when Dick returned and briefed us. We would be leaving for Beale the next morning 24th at 12:30 pm on a tanker.

The next morning, we are bussed to the KC135 tanker. There was a long line of senior ranked officers standing at the loading stairs. I bet every General and Colonel on the base was there. I'm wondering what this was for? We exit the bus and walk past the line. They are shaking our hands telling us to have a safe trip home. We get on the plane still wondering what had just happened. It turns out the next bus carried a Mather crew who had been shot down and recovered. The senior officers were there to send them home! We took off at 12:30 pm and arrived back at Beale at 9:30 am. On the same day (24th). We landed before we took off.

We were on Guam the 23rd and 24th. Each day dozens of B52D and G took off on Linebacker II combat sorties. Being billeted off base I didn't see any aircraft activity on the base. Over all, Linebacker II lasted 11 days, 15 B52s (one from Beale) were shot down, and 33 crewmembers died or were missing. Sometimes I feel like I missed the party, but on the other hand, it was dangerous party.

Aircraft Commander

Back at Beale we resumed alert and transition training. The Wing Commander (Col B) flew with us one day. He sat in the IP seat on the flight. Dick finished air refueling and asked me if I wanted to try it. I said, "Yes!" (Recall that I had lots of air refueling time at Castle, AFB.) The Wing Commander was totally impressed. Col B had me entered into the Pilot Upgrade Program. I was selected for upgrade when I had about 1000 hours.

Kincheloe Assignment, not.

The Strategic Air Command moved crewmembers to another base/unit every three years. There was also the concept of north/south base assignments. Crewmembers moved north/south to share the punishment of cold/snow in the north.

So, my first assignment after pilot training was to B52G at Beale AFB (a southern base). But I was there from 1970 to 1976, or 6 years. So, given the 3-year policy, how did that happen?

There is a saying, "It's better to be lucky than good." Well, I'm just lucky! In 1973 I received orders to Kincheloe AFB, Sault Ste Marie, MI. (I had paper orders in my hand.) However, in 1973 the Air Force ran out of funds to move people so all moves were stopped.

I'm in the bomber/tanker alert facility on nuclear alert and an out-processing Military Personnel Center airman called me. He says, "You have orders to Kincheloe so you need to come down to the Personnel Center and go through out processing." I respond, telling him, "I've read in Air Force Times that the AF has run out of funding so nobody is moving." So, I answer, "No, I don't need to out process." The call ends.

A few days later he calls a 2nd time and we have the same conversation. He calls the third time. Same discussion, but this time I'm thinking I got him now and tell him, "Ok, I'll go through out processing with you but please tell the finance folks that I will be taking advance funding on my move." He says, "No, because they know you're not going anywhere and so they won't give you any money." Frustrated I answer, "Please don't call me again!"

A few days later my orders to Kincheloe were rescinded. Apparently, this event reset my 3-year clock and I spend 3 more years at Beale! I finally left Beale when the B52 unit was deactivated and closed down in 1976. The U2s from Davis-Monthan moved in to replace us.

In retrospect, this was great for me. Had I moved to Kincheloe I would have been a new guy there and would have had to compete against local performers on positions and Officer Effectiveness Reports. Remaining at Beale, I was there so long that I was the senior officer crewmember in the squadron when the unit closed down.

SOS, not

Through my 30 years in the Air Force, I have made many career decisions that others counseled me against. This was one.

There are three levels of military education. The five-week Squadron Officer School (SOS) was for all young captains. To attend the other two 10-month schools, attendees were selected during promotion boards: Air Command and Staff College (ACSC) was for Majors, and Senior Service School (SSS) for new Colonels.

I received orders to attend SOS in residence at a time I was completing pilot upgrade training from copilot to aircraft commander. Not wanting to interfere with my upgrade, I wrote a letter to the Wing Commander declining the assignment saying I needed to complete upgrade.

I didn't attend SOS, but later I completed it by correspondence. I wonder how many others turned it down. By the way, later through the years, I was selected to attend both ACSC and SSS.

My first crew

At Beale, on 20 April 1973, I completed the B52G pilot check ride. I flew a one local solo sortie as aircraft commander on 24 April. I left on a Bullet Shot deployment on 1 May, riding a contract airliner to Guam.

Preparing for a 12-hour Combat Sortie

I met my crew for the first time. Our first flight together was a combat sortie. On 9 May, I flew as aircraft commander with my first crew dropping bombs in SEA! It was a 13-hour all night flight. It was a long 13-hours because the autopilot didn't work so we hand flew it the entire night. We flew combat sorties every few days.

During the deployment I developed an abscessed tooth. A bulge appeared in my gum under a lower right tooth so I went to the base dental clinic. I was surprised to see a full Colonel dentist come into the room to examine me. He looked in my mouth at the abscess for about 10 seconds. He then turned around toward a counter and picked up an instrument. Turning back toward me I see what I believe to be a plyer like tooth pulling tool. He hasn't said a word yet. Concerned I asked, "What are you going to do?" He says, "I'm going to pull the tooth." Really, I'm thinking, and without Novocain?

I respond, "Wait a minute, you looked at the tooth for a few seconds and now you're going to pull it! It will be gone forever. Isn't there another option?" Apparently, this Captain offended or embarrassed the Colonel because he left the room without another word.

A couple minutes later a Captain dentist came in and took a look. He said, "We can do a root canal." Numbed up, he drilled into the tooth, and removed the root, then he inserted medication for the abscess. He then topped off the drilled-out root canal with a putty like temporary filling. Finally, he told me to come back in a few days to check on the abscess and if all was well, he would finish with a permanent filling.

Coffee and Chicken on 12-hour Combat Sortie

I mention to him that I was scheduled to fly a 12-hour combat sortie the next day. He looked at me for a few seconds, then handed me a small drill bit saying, "There is an air pocket under the temporary filling. If the air pressure changes inflight causes you pain, spin this drill bit between your fingers and drill through the temporary filing until you reach the air pocket to equalize the pressure. Any pain should stop.

I flew the combat sortie with the drill bit in a pencil pocket on my flight suit shoulder. Thankfully, no pain, so no drill. I still have the tooth. But what a bizarre experience.

I ended up with 12 combat sorties on this deployment. That gave me a total of 78 combat sorties dropping just under 5,000 bombs. We completed the combat tour and returned to Beale on 4 August.

Fuel Conservation

The 1973 Arab oil embargo pushed the Air Force into fuel conservation. (I was deployed to Guam so missed it.) A number of operational changes followed:

A formal Fuel Conservation program was put in place. It was usually an additional duty for a copilot. The program was inspected and evaluated by the SAC Inspector General.

The low-level training routes were called Oil Burner (OB) routes. The name was changed to Olive Branch (OB). Sure, sounds better. (I know, I'm smiling too).

More attention was made coordinating fuel loads with activity timing. The B52 had maximum weights for different activities such as low level. Many times, we had to increase drag to increase fuel burn by raising airbrakes 4. The increased fuel flow resulted to reduce the aircraft weight to allowable weight for the next activity.

Procedures for descending and landing were changed. After returning to Beale we flew teardrop penetrations to an instrument approach and landing. At 20,000 feet, we would enter the holding pattern over the base navigational aid to be aligned to begin the penetration. The landing gear was lowered, airbrakes 4 selected, and a turn made to track an outbound course. A 240 knot/4000 foot per minute descent was made. Passing 12,000 feet, a 25-degree left bank turn was begun. The plane was leveled off at 4,000 feet. When the inbound approach course was intercepted the speed was decreased to below 220 knots and the flaps lowered for landing.

This meant we flew over the base, then away from it, and finally back to the base. Instead of flying a teardrop penetration from overhead at 20,000 feet, we started flying an enroute descent with engine power way back and directly to intercept the final approach course. We no longer overflew the field, then fly away, then return. This change saved fuel.

Bullet shot, again

I returned to Beale on my first deployment as aircraft commander on 4 August1973. The crew changed only slightly. I picked up a brand-new copilot. After a couple weeks, my crew gets called into a meeting with the squadron commander. We didn't know why. We were sitting in his office and he tells us, "We were tasked to send a crew back to Guam on 15 September and you were selected." I say, "But we just returned two weeks ago!" He said, "A requirement was that it be a solid crew, not 6 crewmembers thrown together." I say, "My radar navigator is enrolled in a college degree program, if he deploys, he won't complete it." (This was important. A degree is important to improve promotion potential.) The conversation was becoming heated. At that moment, I tell my crew, "Crew, leave the room." Out they all scramble and the door closes.

Now it's just me (a Captain) and my squadron commander (a Lt Colonel) boss are in the room. I tell him, "This is BS, the war is over. There is no need for formed crews to deploy. My radar navigator will be screwed over if he goes. When we get deployed the other crews from other bases will be thrown together." I continue, "I'm not buying this. I'm going down to talk to the Wing Commander." He replies, "Don't you understand the chain of command?" My reply, "I do and I'm going through it. I've just talked with you and next I'll be talking with the Wing Commander."

He backs off a bit saying, "Ok, I'll talk to the Wing Commander." My radar navigator was replaced and we deployed.

(Many years later my then "new copilot" reminded me about the squadron commander meeting. He said when the crew left the office and closed the door, they all put an ear to the door and heard the entire conversation. He said that it was the bravest thing he had seen. I'm a Captain arguing with my boss, a Lt Colonel. He said that I was someone he wanted to be associated with because I took care of my people.)

On 15 September 1973, my crew was deployed in the B52G to Guam on 140-day orders. We rode a KC135 tanker to Guam. The South East Asia (SEA) war was winding down. The ramp was saturated with B52Gs and B52Ds. Guam started to transition from the conventional war to support the nuclear Single Integrated Operational Plan (SIOP). My crew was among the first 14 crews to go on nuclear alert on Guam.

The 14 crews were called into a briefing where the DO told us we would certify the SIOP, in 4 hours! I was in shock. Four hours! Absolutely not enough time to adequately prepare! This put me into a total panic! I had certified before, many times. A SIOP certification was a serious formal event. Four hours of SIOP sortie study time as not enough time to learn what was needed and prepare to brief and respond to questions.

My experience was that a SIOP certification occurred after a many days of sortie study. Topics included Command and Control messages and procedures, aircraft performance, route study, air refueling, action points, low level and specific target study. The certification briefing was given to a Colonel and the wing staff from a stage. Each member of the six-man crew stood up individually and briefed a portion of the mission. Then the Colonel and the staff asked questions, lots of questions. It was a serious event. If you passed you were certified, if not, more study.

So, when I was given 4 hours I was in a panic. For 4 hours we frantically looked through the classified sortie bag learning what we could about the sortie. At certification time a Colonel came in and sat down in front of the assembled 14 crews. He talked to us for a few minutes, he then declared that we were all certified! What! No formal briefing, no questions. We had no real understanding of the sortie we were now responsible to fly. So now we were on nuclear alert. And yes, with nuclear weapons. While on alert we studied the sortie and then became confident that we could fly it.

A klaxon (horns) sounding while on alert directed the alert crews to respond to their aircraft, start engines, and copy a coded Emergency Action Message. Once a week the Strategic Air Command conducted a no notice exercise alert. The Anderson klaxons (horns) blew and 14 crews responded to their aircraft and started engines. Simultaneously with the klaxon, a coded Emergency Action Message was being broadcast over the UHF radio. We didn't know if the Klaxon was a real war event or an exercise until a message was received and decoded. This decoded to an exercise.

Gun powder start cartridges were used to quickly start the B52 engines. The cartridges burned sending gas through the engines' starter that spun up the engine RPM to just above what was needed to start. The cartridges emitted a massive amount of white smoke. During this first exercise one of the pilots thought all the smoke was a fire and called an emergency termination of the exercise. Wrong!

The next day all 14 crews were directed to the flight line to watch a B52 cartridge start. Fourteen alert trucks and crews formed in a semi-circle around the front of a B52. The engines were started with cartridges so everyone could see the smoke during a cartridge start.

Bangkok Trip

We sat a 7-day nuclear alert tour. Coming off alert on October 13th, Tom (copilot), and Rick (the gunner), and I took a Young Tiger KC135 tanker ride that passed through Kadena AB, Okinawa, then continued on to Utapao (UT), Thailand. This trip occurred 13-17 October. At UT, we hired a baht bus. A baht bus was like a cab. Baht is the Thai currency. Back then, 1 Baht was the worth about 5 US cents.

The bus was actually a small truck with sides and a cover. We paid the baht fee and sat in the back of the truck for the ride from Utapao AB to Bangkok. We stayed in the Florida Hotel. We weren't able to do any sightseeing. On the 14th anti-government riots had started. That morning the hotel staff didn't show up. Looking out across the city from my hotel room window, I saw smoke rising from various areas across the city. It appeared that off in the distance helicopters might be shooting down toward the ground! The police had disappeared. The traffic circle next to the hotel had become jammed to a stop with vehicles. People were on the street requesting blood donations!

The three of us left the Florida Hotel and walked crossed the street to the Choupia Hotel. I'm leaning over the

Riding a Baht Bus to Bangkok

counter writing a post card and hear a guy talking next to me. I know the voice! I look up. I went to pilot training with John. John and his wife Susie were in the lobby. John was assigned in Taiwan flying the C130. We agreed that we needed to get out of the city and decided to meet in Pattaya beach for dinner that night.

The three of us left the hotel and haggled with a baht bus driver for the fare to drive us to Pattaya beach. We told the driver that his price was too high and started to walk away. About that time there was some automatic gun fire just down the street. We did a 180 turn, gave the baht bus driver the money he requested and away we went. As we drove out of Bangkok, we passed a military armored vehicle column driving into the city.

Jammed Traffic Circle in Bangkok

That night, as planned, we met John and Susie at the Pattaya beach restaurant for dinner. We had a great reunion. During the meal my copilot, Tom, became a bit tipsy! He didn't think that his meal was cooked correctly and sent it back to be redone. I'm convinced the waiter staff put something in his food because later he became quite ill. That night we stayed at a US Embassy beach house cabin.

The next day we toured around the area. That night we ended up in a bar along the beach. Except for the three of us, the bar was completely full of Thais. Tom and I stood at the bar watching Rick (our gunner) out on the big dance floor just cutting a rug. Rick stood about 6 feet 4 inches tall and about 230 pounds. He was at least a foot taller than anyone else on the dance floor, the sole tall man in a sea of short people! A few minutes later Rick was standing next to us at the bar. Rick passes out falling forward over the bar. He slid off the bar and ended up on the floor. I swear he left teeth marks across the bar!

What happen next was truly amazing. As soon as Rick hit the floor, a mass of Thais converged on him and picked him up. Dozens of hands lifted him overhead and carried the 230-pound man out of the bar. They loaded him into the back of a baht bus. It was like the Lilliputians and Gulliver! Off we went to the beach cabin for the night.

The next day we rode the Young Tiger KC135 back to Guam arriving on the 17th. We looked pretty scruffy and probably smelled fairly nasty. Between the three of us we had just carried one small bag with all our stuff in it. Going through US Customs, the Customs guy asked, "Whose bag is this?" We honestly tried to explain that it contained all our stuff. He asked the question a couple more times. Not a good or right answer. He apparently wanted "someone" to claim it. They separated us, searched the bag, and body searched each of us! How fun was that!

Our Guam tour was cut short when the 1973, October War started in the Middle East. We redeployed to the US on what I call, "The trip from hell."

Trip from Hell

What is fatigue? Fatigue is the condition of being very tired as a result of physical or mental exertion. During peacetime, there was a requirement that aircrews have 12 hours (8 hours uninterrupted) of crew rest prior to the scheduled show time. Show Time was 2½ to 4 hours before the scheduled takeoff time. The normal crew Duty Day was 12 hours, or 16 for Higher Headquarters Missions (24 if augmented with extra crewmembers). There was no transition (practice approach/landings) after 12 hours.

The Cold War began after WWII with the East versus the West. SAC was always on a war footing which resulted in 30% of the bomber force on alert at bases throughout the USA. There were nuclear weapons loaded aircraft that were assigned for specific targets. We lived in alert facilities by our aircraft near the end of runways. Alert tours lasted for up to seven days at a time. On the average, each crew was on nuclear alert about 10 days each month. We were tasked with the requirement to be able to take off within 15 minutes of notice. ICBM missile flight time from the USSR was 30 minutes. This response time would allow us to takeoff and leave the base area before inbound Soviet nuclear missiles could destroy them.

In 1973, the escalation of world events and tension put us into a dangerous situation. The Yom Kipper War in the Mideast was in progress during the period 6-25 October 1973. On the 24th, the Soviet Union put a few airborne divisions on alert. This action was seen by the US as an apparent preparation for Soviet forces to be inserted into the Middle East region. The US reaction was to put our nuclear alert crews into restricted alert and started a SAC force wide nuclear generation of all SAC forces. Stateside base deployed B52Gs on Guam would be redeployed back to the US and generated to nuclear alert status.

On Thursday afternoon the 25th, I was scheduled to fly as an instructor with a newly arrived crew on a local Guam training flight. The rest of my crew was off for the day.

I arrived at the aircraft with the new crew at 3:30 pm on the ramp to begin the aircraft preflight. We were soon informed that all flying was cancelled for the day. Why? We didn't know. Later that day we heard of the rumors of the increased nuclear alert force posture. Alert force crews' freedom of movement was restricted. There were 14 B52s on the Guam ramp on nuclear alert. All non-alert crewmembers were put into crew rest at 4 pm. Nothing much was heard of events throughout the day, that is, until after dark.

Thursday night was typical with low rainy scud clouds blowing over Guam. Around 9 pm the alert klaxons (horns) sounded! I came out of my second-floor room and stood at the railing watching all 14 alert crews run to their alert vehicles then drive rapidly down to the aircraft parking ramp to their respective aircraft. Soon the alert aircraft engines were starting.

When in an increased posture, there were no provisions for exercise alerts so I believed this to be a real-world event! For a few brief minutes I actually thought, in total disbelief, that nuclear war had started and that Soviet Union missiles were enroute to the island! Yep, I thought I was going to die in a few minutes! Wow! Well, the planes didn't taxi and takeoff. Crews began to shut down the aircraft engines.

The Bomb Squadron commander walked by and said it was an inadvertent klaxon. Perhaps a short caused by the rain! Who knows? The alert crews re-cocked their aircraft to resume alert status and returned to the alert facility area. Whew! No missile strike, so we would live to fly again!

A few minutes later, about 10 pm, the Bomb Squadron commander came by a second time. He told me that in the morning my crew would be in the first 3 B52G cell to redeploy to the US. Once back in the states we would be generated to nuclear alert status. I was given a 1 am assembly and briefing time for a 5 am takeoff! Crew rest, what's that? I was technically put into crew rest at 4 pm, but I had been up since 7 am that morning. Now no sleep tonight! I had a real scramble around the base to find my crew, pack everyone up, and be ready to meet the 12:30 am bus time.

The three crews assembled at the 1 am briefing. The current operations shop was in total chaos with lots of commotion. Things were happening very fast and they appeared to be overwhelmed. All of our flight planning information and documents I had been given showed Mather AFB in Sacramento, CA, as our destination. The other two aircraft were heading to Wurtsmith AFB, MI. The briefer kept telling me that we were going with the other two aircraft to Wurtsmith AFB. I kept saying Mather AFB! It was sort of an ongoing argument. We actually left the briefing room with the briefer yelling at us that we were going to Wurtsmith!

We arrived on the ramp at the designated aircraft parking location around 3:30 am, but there wasn't an aircraft parked there! It was being towed up the ramp toward us. There was a problem with an engine starter and the maintenance folks still needed to work it. Most of my crew went to base operations, the snack bar, and the class 6 store (buy liquor to carry to the US). The copilot stayed with me at the plane waiting for the plane to be repaired. Obviously, we didn't make our 5 am takeoff time. At 4:30 am the other two aircraft taxied and took off at 5 am.

The second cell of three B52s was scheduled to depart at 7 am. Our engine was eventually fixed, we completed the aircraft preflight, and sat in the plane ready to start engines.

The operations coordinator radio call sign was "Charlie." He was a Colonel located on a slight hill overlooking the parking ramp and runways. Charlie broadcasts over the UHF radio, "Everyone ready to start engines, start," then, after a few minutes, "Everyone ready to taxi, taxi." Three B52s taxied past us. We taxied out behind them as the fourth aircraft.

The redeployment operation was obviously in an extreme "getter done" mode! As we sat number four waiting behind the others at the runway hold line, Charlie now asked, "What is your tail number, aircraft commander's last name, and where are you going!" The three aircraft in front of us were going to Wurtsmith, we to Mather. With that information Charlie told the 4-ship cell to contact the control tower for takeoff clearance.

We took off one minute after the number three aircraft. The time was 7:18 am on 26 October. At this point we have already been up over 24 hours. In about 13 hours we would be arriving at Mather AFB and almost home! Beale AFB was just a 45-mile drive to the north.

Not long after takeoff the first indication of crew fatigue surfaced. As we climbed through 10,000 feet, the 10,000-foot checklist was completed. It called for the cabin pressure be checked. I looked up at the cabin altimeter (cabin pressure). Our cabin wasn't pressurizing! The cabin was at 10,000 and rising with the aircraft altitude. It should have been somewhere around 4,000 feet. We called the cell leader and told them we were leveling off to work the problem.

The other three aircraft continued to climb and accelerate away from us as we leveled off. After a short investigation of the pressurization system issue the navigator calls over the interphone that he found the problem. There is a small hatch downstairs behind the two-man navigator position. The hatch can be used to crawl back into the forward landing gear and bomb bay area. It wasn't closed and locked! It was swinging open. It was an obvious miss on the navigator team preflight checks! Fatigue!

The hatch was closed and locked. The aircraft cabin pressurized. We once again began the 280-knot climb. By now we were many miles behind the other three aircraft. As we climbed and neared our assigned latitude I looked over at the copilot. Tom was asleep. His helmeted head hung forward. His oxygen mask was connected to the left side of his helmet. As the aircraft went through minor turbulence his head and mask would swing side to side. Fatigue!

I let him sleep. I knew we were facing a long tough flight and any rest he got now would help later. Or so I hoped.

We leveled off around 33,000 feet. The other three aircraft were well out of sight ahead. We needed to join them. The cell leader slowed the three aircraft 30 knots and we accelerated 30 knots. We expected to catch them in about 15 minutes. It was a clear morning and I watched ahead for the other aircraft. The copilot is still asleep. After about 15 minutes I still couldn't visually see the other aircraft.

I ask the radar navigator over the intercom. "Radar, pilot, do you see the cell on the radar yet?" I get no response. "Navigator, pilot, do you see the cell on the radar yet?" Nothing, no response! I call, "EW? Guns? You guys up?" I get no answer! I look over at the copilot, still asleep.

We haven't been airborne an hour yet and I'm the only person awake in the plane. Only about 13 more hours to go! Fatigue! I call for the Radar again and this time he answers. "Do you see the others on your radar?" "No," he says, he sees nothing. The radar navigator calls the cell lead navigator and asks him to read his coordinate counters. The radar checks lead numbers against ours and discovers we have passed the other three aircraft! We are ahead of the other three aircraft. That's why they couldn't be seen on the forward-looking radar. Fatigue!

So now we slow 30 knots and they accelerate 30. After about 5 minutes I see them moving forward off to our left a few miles. I call, "Tally Ho" and as they move ahead of us I maneuver left into a trail position behind number three. We were stacked up 500 feet altitude above number three. Finally, the cell is formed and we all accelerated to enroute cruise speed of Mach .77, with a tail wind it's over 8 miles a minute.

It was a tough start for a flight. We were making serious mistakes and errors. Obviously were all tired and fatigued. The scary part was that we were now settling into the most boring part of a flight, hours and hours, of high-altitude cruise. We had hours of boring drone time ahead of us. I was glad we were the last aircraft. We weren't sandwiched between aircraft and thus had more room for error. I wondered what the fatigue situation was inside the other three aircraft. I knew our situation was critical.

We flew east through the time zone changes to the east as the sun sped west. It was a short duration daylight and it quickly became night again.

There were times when we were all asleep! I tried taking turns with the copilot monitoring the aircraft so we could take turns sleeping. When it was my turn to monitor the aircraft, I'd fall asleep, we were all asleep, everyone! The plane would hit some turbulence and shake. I'd wake up. Way out in front, miles ahead, the rotating beacons of the other aircraft could barely be seen. I'd increase the power slightly to catch up with them. Then I'd pass out again. Next, I would awake again to see rotating beacons of the other aircraft in my face meaning we were really close behind the other aircraft, I'd pull power in order to move back. Then fall asleep again. And so, this went on for an hour or two! It was absolutely terrifying! I don't know what was happening when it was the copilots' turn to fly.

We eventually approached the Oregon coastline. The four-plane cell changed from HF to UHF radios and contacted Washington Center. I'll never forget the air traffic controller's initial call to us, "Welcome home!" It sure sounded great, we were back in the states. The three aircraft in front of us continued east toward Wurtsmith, AFB. We had shared a terrible flight together and I didn't even know who they were. We turned south toward Sacramento. We were now on our own.

It was a clear night as we flew down the west coast. We eventually descended into the Sacramento Valley and contacted the Mather AFB command post.

The entire crew was obviously dangerously tired. It was difficult for me to even focus on the flight instruments. Concentrating on anything was almost impossible. I would shake my head and blink my eyes in an effort to stay awake, focus on, and most importantly, understand what the flight instruments were telling me.

All of the Mather B52Gs had been deployed to Guam. We were flying the first B52G returning to Mather. The Mather folks waiting for us had orders to generate the aircraft returning to nuclear alert status as soon as they could. They requested that we not deploy the drag chute after landing. The drag chute is normally deployed after landing to help slow the plane. Not deploying it would save them at least 30 minutes reconfiguring the aircraft to nuclear alert status.

I knew we were in a dangerous physical condition so I told the copilot that we would use the drag chute on landing. I wasn't doing anything out of the ordinary. We landed at Mather AFB on 26 October at 2:17 am. The drag chute was deployed. We taxied onto the parking ramp and shut down the engines. We were the first B52G to return to Mather, AFB. I was still strapped in my ejection seat when a US Customs official came up into the cockpit. I signed a bunch of paperwork for him. I had no idea of their significance. I then followed him down and out of the aircraft.

I wasn't in much of a thinking mode. All I wanted was to get the plane unloaded and go somewhere to sleep. Standing bent over on the ramp at crew hatch under the aircraft a crew suitcase was passed down to me. I swung around and shoved it into the arms of the guy behind me. I assumed that the hand I saw was reaching out toward me was for the suitcase intending to help us unload.

Turns out it was the Wing Commander! He was receiving his first aircraft back and had apparently stuck out his hand to shake mine. Instead of a hand shake, he got a chest full suitcase. I don't recall his name. I don't even know what he did with the bag. I think he passed it to someone else and left the flight line. I was staggering with fatigue.

Because of the flight timing and time zone changes, we landed at Sacramento before we took off from Guam! We departed Guam at 7:18 am in the morning on the 26th. We landed at Mather at 2:17 am on the 26th.

A couple of days later my crew was on B52G nuclear alert at Beale AFB. As for the Middle East situation. The Soviets saw the US reaction and did not activate the airborne divisions and did not deploy troops to the Middle East. The 1973 October War finally ended. World tension decreased and we returned to normal alert status at Beale.

A few months later, my crew was selected for one final Arc Light (B52D) deployment. We spent 11 Days at Castle AFB, CA, checking out in the B52D, practicing 3 plane formation, and conventional bombing procedures. Been there, done that.

From Castle, we rode a KC135 through March AFB, CA, then onto Hickam AFB, Hawaii where an engine change was accomplished. The next day we flew to Guam, refueled, then continued on to Utapao, Thailand.

We were there from 15 January through 26 Feb 1974. During that time, we flew 5 crew conventional training sorties. Since I was an experienced pilot in conventional operations, I also flew on 2 training flights with new inexperienced crews.

On 1 February, I had a treat. One of my Wisconsin State University college ROTC buddies (Bob) was there crewing a KC135 tanker. I rode along on a 3-hour flight with him.

9 February was a tragic day. A crew back at Beale AFB ran off the runway during a takeoff attempt. All 6 crewmembers and 3 evaluators were killed. They were friends of mine.

The war in South East Asia was over. North and

South Vietnam had signed a peace treaty ending the war. Our flight operations changed. There were no combat sorties.

A KC135 flew us to Guam on 26 February (we stayed on Guam until 19 April.) A requirement before sitting alert with nuclear weapons we needed to demonstrate we knew the mission. The next day we studied, and then certified the nuclear war mission. We accomplished a certification by briefing a nuclear mission to a Colonel, then successfully answering his and staff questions.

When we were on nuclear alert 6 of us slept in one room. One member snored. He would start to snore, copilot Tom would call out, "Rick" which would wake me. I would throw a paperback book at the snorer and he went silent, for awhile. Then it would occur again. In the morning there were 5-10 paperback books around the offender's bed.

Life on Guam was interesting. When not on alert, the 6-man crew lived in a two-adjoining room (three beds to a side) with a walk-through adjoining bathroom/shower.

One block away was an open-air outdoor theater. We called it the Drip Dry. It provided something to do. We watched a lot of movies there. The discussion sounded like this, (Tom) "Rick, do you want to go to the movies tonight?" (Me) "What's playing? (Tom) "I don't know." (Me) "Sounds good, let's go." We sat many nights wearing rain coats and under umbrellas.

Sometimes we got into mischief. We built tennis ball bazookas that shot tennis balls. Simply tape tennis ball tubes together. Make a hole at the closed end of the tube, squirt in a little lighter fluid and roll a tennis ball into the tube then light the hole. Boom-a tennis ball bazooka war between buildings was on!

Sometimes things could get rowdy. In all my deployments, I only had one interaction with the security police. It was late, I was asleep, the phone rings. The call was from the Anderson Base jail (the Brig) asking me to come down, identify, and pick up my gunner. I get there to see his face looking through the jail bars with hands holding the bars. He has two black eyes. I tell the sky cop, yes, that's my gunner, they release him, and we leave.

He was arrested for fighting with a few navy personnel. It turns out that Rick was out with a number of other young gunners. They were in a bar and got into an argument with some navy folks. It became physical. When he was squared off with a navy guy another one came up behind him and pulled his ankles (feet) back out from under him. He hit the floor where he got punched in the face.

Fight over, his companions drove him to the Anderson AFB, hospital emergency room and dropped him off. They returned to the bar and got into another fight with the same navy men. The police arrived and arrested everyone. The police then went to the base hospital, arrested Rick, and took him to the Brig.

We were leaving Guam in a few days. The Anderson legal office gave me a sealed package about Ricks arrest and the charges against him. They told me to give it to the Beale Legal office.

On Guam, we flew three nuclear war crew training sorties and sat alert on 3 seven day long alert periods. The plan was for us to redeploy to Beale on April 15th but as we know, plans don't always work out.

We departed Guam on 15 April, planned to fly two Ds to March AFB, CA. Two KC135 tankers had taken off earlier and were orbiting waiting for us at 26,000 & 26,500 feet. Our two B52Ds leveled off at 25,000 and 25,500 feet. We needed their fuel to fly to the states.

For refueling, the formation positioning changes. Instead of being in trail, the following aircraft move to the right on about a 60-degree line approximately two miles away from lead. The tankers rolled out ahead and above us and slow to 255 knots refueling speed. We chased them at 280 knots until 1 mile, then we slowed to 270 knots and began a climb toward the tankers. We were in solid clouds and closed on them using our aircraft radar. The other bomber was also closing on the second tanker with radar. The next target was 265 knots at half mile and 500 feet below the tanker. At the half mile position, to continue up we needed to visually see the tanker.

At 500 feet neither bomber could see its tanker. In fact, none of the four aircraft could see anyone else. I suggested that the tankers climb 1,000, then we climbed 1,000 feet. Still no visuals at half a mile. We climbed again without success. Chasing the tankers, we flew to the planned end air refuel point. The tankers now couldn't off load our needed fuel and have enough for them to make it back to Guam.

We had to abort the flight and return to Guam. There were four aircraft in close proximity who couldn't see one another. I called for the B52s to descend 1,000 feet. Once leveled off there, and we had separation from the tankers, I had the tankers turn back toward Guam. Once they were heading to Guam, I turned the B52s back toward Guam. We landed after a 6.3-hour flight.

Four days later, on 19 April, we departed Guam again. This time we accomplished the needed air fueling and flew 13.3 hours to March AFB, CA.

At March we planned to remain overnight. We each travelled with a number of bags. One of mine was called B4 bag. It was square shaped canvas bag with a zipper over the top. We unloaded in the parking lot in front of our rooms. I unzipped the bag, started to reach in but then I heard hissing (snake?) I quickly pulled my hand out and backed away. Keep in mind, there are lot of snakes in Thailand and Guam. I found a long stick on the ground and poked around. I found the culprit. My battery powered radio had turned on and was generating static hissing! The next day, a KC135 flew us up to Beale AFB.

BTW, I ripped up the sealed legal package I was told to give to the Beale Legal office. I figured my gunner had been punished enough. Rick was my gunner, I take care of my people.

Rapid Decompression at 3,000 Feet

Here is some background information. In 1969 I attended Air Force Pilot Training at Randolph AFB, San Antonio, TX. One required training course was physiological training. The purpose of the training was to familiarize the crewmember about the physiological characteristics of flight and how they interact with the human body.

One of the subjects was hypoxia. We learned how changing air pressure impacted on the way oxygen entered the body. As air pressure around us decreased, less oxygen is able to pass into the body through the lungs and eventually we become hypoxic (a state in which there is too little oxygen in the blood), then unconscious.

From classroom academics we moved to an altitude chamber where we experienced it. There were 12 students seated with backs against the two chamber sides. A technician was in the chamber with us. The chamber operator could be seen through a small window at the end of the chamber. There was a Flight Surgeon outside in case a medical issue developed. Each student had their own oxygen system, controls, and mask. We pre-oxygenated with 100% oxygen for 30 minutes to breath off nitrogen from our bodies before the pressure was changed. The reason was to reduce the chance of getting the bends. Each person has slightly different symptoms as they experience pressure changes and become hypoxic.

I had passed my TUC. The attendant put me back on Oxygen

After 30 minutes the chamber operator pumped the air out reducing the air pressure in the chamber. This simulates the thinner atmosphere at high altitude. After the chamber has been mechanically elevated to a high altitude each trainee has the opportunity to remove the oxygen mask so they can observe their individual symptoms of hypoxia. Finger nails get blue, breathing becomes deeper and labored, vision becomes dimmer, and finally the individual gets to the point where they may be awake but can't help themselves. This moment is called Time of Useful Consciousness (TUC). If the aircraft pressure was lost there was limited time to get on 100% oxygen.

A Rapid Decompression (RD) is an instantaneously loss of cabin pressure. A RD reduces by half the amount of time a person has to react and put on the oxygen mask. Air pressure decreases instantaneously: there is a loud bang, a sudden momentary cold mist like fog forms, then disappears, and the air rushes out of the lungs.

I attended refresher training in a chamber every 3 year. Each time I attended this training event I noted that one of my symptoms was a strange sensation in my mouth when I swallowed. I'm convinced my strange swallow sensation saved my life.

Swallow, try it, you close your mouth and kind of suck, then swallow. I noticed that pressure in my mouth was different in a pressure situation versus when I'm on the ground.

On 6 May, 1976, I flew a B52G from Beale AFB, CA, to Tinker AFB, OK, to turn the plane into Depot Maintenance. We would pick up a B52G from Depot Maintenance that had just completed 4 months of heavy inspection, modification, and maintenance.

After we turned the Beale plane over to depot maintenance, we preflighted the B52G that we would fly back to Beale, started engines, and taxied out to the runway hold line. It all went surprisingly smooth considering it had been on the ground for 4 months, torn apart, and reassembled.

In the B52, the takeoff was made with the pressurization system OFF and with RAM air selected. The pilot turns pressurization system ON by selecting 7.45 PSI differential after the landing gear and flaps are retracted.

According to the climb checklist, the first time the crew cabin altitude (pressure) was checked was climbing through 10,000 feet. The purpose was to insure the cabin was pressurizing correctly before climbing to higher altitudes. There is a cabin pressure gage located just above the front windows in a line with the 8 engine oil pressure gages. It measured and displayed what the air pressure was in the cockpit.

If the pressurization system was working correctly, it would indicate a pressure well below 10,000 feet, probably around 4,000 feet. This indicated the crew compartment was pressurizing as we climbed. If it was 10,000 then we weren't pressurizing.

After takeoff the landing gear came up and the flaps retracted normally. I selected the pressurization system to ON, 7.45 PSI. The airspeed was increased to climb speed 280 knots.

So far, so good. But then I swallowed. I instantly realized the swallow sensation that I had in the altitude chamber training….one of my pressure problem symptoms!

How could that be? We were only at 3,000 feet above the ground?

I looked up at the cabin pressure gauge. The needle read below zero, below sea level, off the scale and rapidly descending. What? It should have been slightly below 3,000 feet and slowly climbing. I had never seen or heard about this situation before. For some reason our cockpit was over pressurizing, the pressure inside was greater than outside! Over pressurizing, at some point the crew cabin would probably explode!

I wasn't sure what to do but I knew I needed to stop the cockpit pressurization immediately so I reached down and slapped the EMERGENCY DUMP switch to DUMP. The DUMP position immediately releases (dumped) all the pressure from the crew cockpit to match the external air pressure.

The moment I selected DUMP we had a Rapid Decompression: the loud bang, the cold mist appeared/disappeared, and the big pressure change. The cabin pressure gage needle popped up from somewhere below sea level off scale up to about 3,000 feet above the ground.

We had a Rapid Decompression (RD) at 3,000 feet! Normally an RD would be expected when the plane was at high altitude 35,000 feet, for example, with a cabin pressure of around 8,000 feet. An RD would instantaneously cause the cabin pressure to go to 35,000.

I did a quick crew check over the intercom to check the crew status. The pressure change hurt but we were all ok. I levelled off at 10,000 feet. We called Beale on the HF radio and discussed the event with our maintenance personnel. They suggested I try it again. I said, "No way!"

Without cabin pressurization, I decided to fly back to Beale in RAM at 18,000 feet. That meant at 18,000 feet outside, the pressure in the cabin was also 18,000 feet. We had our helmets and oxygen masks on breathing oxygen.

There is always something else. We experienced clouds and icing over the Rocky Mountains so I turned ON the aircraft anti-ice system. About every minute the pressure would suddenly spike then drop. Pretty annoying after awhile. Initially we didn't know what the cause was. The B52 has an extensive flight manual. I studied the manual constantly and was tested on it routinely. But there are nuances in the manual that seldom occur. I thumbed through it and found the reference that said when the anti-icing is ON and the RAM air scoop is out, once a minute the ram air scoop would slam in then back out. The purpose was to knock any ice accumulation off the RAM air scoop. It was a long uncomfortable trip back to Beale where we made an uneventful landing.

Analysis: My swallow and recognition that something was not right saved the plane and probably the crew. If I had not detected an issue and selected DUMP the pressurized cabin would have over pressurized and the crew compartment would have blown-up!

This KC135 was Over Pressurized during a Ground Pressure Check

When pressurization is turned on, engine bleed air from all 8 engines flows unchecked into the cockpit. The cockpit pressure differential is regulated by three outflow valves. These valves adjust constantly to keep the cockpit pressure at the desired selected level, 7.45 PSI. None of the outflow valves are visible to the crew. Two valves are behind the pilot's circuit breaker panel. The third is near the floor forward of the Navigator team. Checking them during the aircraft preflight is not a crew function.

Post flight, the Beale maintenance folks found **all three outflow valves were in the locked closed test position**. Bleed air was jamming into the cockpit, no air was being released so we over pressurized!

Someone once told me that a B52 coming out of Depot Maintenance was test flown locally 3 times before it was turned over to a SAC unit.

I have wondered over time. Did a mechanic make a mistake, or did someone try to kill us?

Satellite Alert Hill AFB

Back in the Cold War years nuclear bomb loaded B52s sat alert status at SAC B52 bases all around the country. I sat alert at Beale AFB, Yuba City, CA. To complicate Soviet Union targeting, satellite alert was established to disperse the bomber force. B52 units deployed B52s to satellite alert bases across the country. Missile flight time from the Soviet Union was 30 minutes. We were tasked to be able to takeoff within 15 minutes, day or night.

I sat many 7-day alert tours at Beale during my 6 years there. I also sat satellite alert at Mountain Home AFB, Idaho, and later at Hill AFB, Utah. We travelled to the satellite alert locations two ways, riding a C130 cargo plane or by flying in a B52. Usually, we rode a Hamilton AFB C130 aircraft traveling to/from our satellite alert locations. Crews and maintenance personnel arrived onboard along with our baggage.

We used heavy wheeled electric generators and a heavy-duty air start cart to support the alert aircraft. These were Aircraft Ground Equipment. Occasionally this equipment needed to be transported. It would be secured down the middle of the C130. The fuel tanks would be topped off for the flight so there wouldn't be any fuel fumes in the tanks. Even though there weren't any airline style seats, the C130 was fairly comfortable ride. We sat on a canvas bench seat that ran down along both sides of the cargo compartment.

One memorable event occurred on a return trip from Hill AFB to Beale AFB. I was seated toward the aft end of the cargo compartment. The center was filled with a line of Aircraft Ground Equipment, all topped off with fuel. The C130 lands. As we slowed and turned off the runway, I looked forward toward the cockpit entrance. All I saw was smoke! A thick solid wall of smoke. I couldn't see the front of the cargo area.

Concerned about fire, I unstrapped and moved to the aft jump door. These opened by reaching the handle at the bottom of the door and pulling the door up. I squatted there holding the opening handle for a few seconds watching the situation. Even though the plane was still taxiing slowly, if I had seen any other fire indication, I was prepared to open the door and jump out onto the taxiway.

The loadmaster appeared out of the smoke bank and saw me. He ran toward me waving his arms yelling, "No, no, no, no." Some electrical equipment below the flight deck fried and caused the dense smoke.

The B52 aircraft alert changed out every 45 days so sometimes we flew a B52 into/out of the satellite location on a training flight returning to Beale AFB. This first flight after 45 days on alert was called the First Sortie after Ground Alert. Any maintenance issues that occurred on this first flight were analyzed to see what impact would have been had it launched on a nuclear war sortie.

I deployed as a copilot sitting alert at Mountain Home, Idaho. It was a TAC F111 fighter base. Once after our 7-day alert tour ended we taxied a B52 out for takeoff and passed rows of F111s without engines. It was the President Carter years. Not much of a military budget so no engines.

After I up graded to aircraft commander, I deployed to Hill AFB, Provo, Utah, a few times to sit nuclear alert. It was a great location! Being away from the Bomb Wing staff was great; no ground training or being bothered by them. We had lots of sit around quiet time. I would sit outside and enjoy a beautiful view of the snow-covered mountains.

We flew a B52 into Hill AFB a number of times. The first time, we flew an instrument penetration, descent, and approach. Well, at least we tried. Hill was a fighter base. We didn't realize the penetration descent rate was designed for a fighter. A B52 descent was designed for 4,000 feet a minute. This fighter descent required 7-8,000 feet per minute. Half way down I realized we couldn't make the altitude requirements of the published penetration and had to ask approach control for a 360 descending turn to lose more altitude. How embarrassing is that!

We finally got lower and we ask approach control for a few practice instrument approaches before landing. Weather was actually great, temperature warm, and winds calm. I decide to screw with the crew.

During each of the three approaches and patterns the copilot and I made numerous BS comments over the intercom about how cold it was, how bad the blowing snow looked. After the third approach we landed. When we deplaned into a beautiful day, the navigator team was wearing their cold weather gear, including parkas and mukluks. Remember the nav team had no windows in their lower deck compartment and obviously had not listened to the arrival weather broadcast. We all had a great laugh.

Another time we flew a B52 into Hill, AFB. It got ugly. We had an Air-to-Ground Missile (AGM28) under each wing. Each AGM28 flew with a jet engine. There were no weapons installed for the ferry flight. Nuclear weapons were loaded after landing for alert.

The missile engine controls were on the copilot side panel. The missile engines were supposed to be changed from cruise power to idle, then shut down before landing. The copilot didn't make the change to missile engine shut down! I didn't catch the error. I landed with the B52 eight engines in idle. Unknown to me, the two AGM28 missile engines were in cruise power setting! It was a real challenge to stop the aircraft with drag chute and heavy use of brakes! I managed to get the plane slowed down and we turned off the runway at the end of the runway. We then figured out the problem and shutdown the missiles.

Once we flew in with the new B52 but wouldn't go on alert for another day. We stayed in officers billeting. It was Sunday and we had no transportation to drive anywhere for dinner. The only place open to eat on base was at the Officers Club. My copilot and I walked over to the club. We go in and sat down in the dining room. The manager comes over and says, "We have a requirement to wear a coat and tie for dinner." I explain, "We just arrived to sit SAC alert, had no other options to eat on base, and we didn't carry coats and ties."

He smiled, nodded, and said, "Follow me." He took us to a cloakroom with a rack of hanging coats and ties, he says, "We have this to take care of your situation." I respond, "Thanks." He leaves.

I tell my copilot, "Time to screw with Tactical Air Command." We were in the Strategic Air Command. This was TAC country. SAC flew B52s, TAC flew fighters. I'm talking totally different cultures so it's time to screw with them. I pick out a jacket that is way too small for me. I jam into it, sleeves up at my forearm, can't button the front. My copilot put on a jacket way too big for him. It hung on him with sleeves hanging past his hands. We both picked out ridiculous looking ties.

Now decked out with the required official coat and tie, we proceeded out into the dining room, looking totally ridiculous, and sat down in the middle of the room surrounded by all the TAC folks. Nobody approached us. There was some commotion, but after all, we wore coats and ties.

Alert at Hill was great! I always enjoyed my alert time there. Seven days away from the Wing staff.

The Strategic Air Command conducted a no notice alert exercise once a week, once each alert tour. The purpose was to exercise the entire system, the command and control network, command centers, and train the crews and ground maintenance personnel. And just as important, it also showed the Soviet Union we could beat their missiles and launch our nuclear loaded bombers before they could destroy them.

Alert klaxons sounded and an Emergency Action Message (EAM) was transmitted. Crew response to a klaxon was to immediately go to the alert aircraft, start engines, copy and decode the Emergency Action Message.

Responding alert crews didn't know if the klaxon was an exercise or real world "go to war" message until it was copied and decoded. Alert were tasked to be able to takeoff in 15 minutes, longer and Soviet Union nuclear missiles could kill them.

The klaxon sounded and we raced to our B52 from the alert facility. The navigator team copied and decoded the message. It was a SAC engine start exercise. We sat there ready to start engines waiting for the copilot to arrive. He was on the other side of the runway at the base movie theater. To let the crew chief know what was going on I say over the intercom, "Ground, pilot, we are ready to start. Just waiting for the copilot." "Roger," he responds.

The crew chief knew how important it was for us to make our timing. Unknown to me, he does some coordination on the ground. A minute later the crew chief says, "Pilot, ground, I have an engine run qualified maintenance sergeant down here that can come up and help you start engines while waiting for the copilot."

This would certainly improve our status. However, this was a new situation for me! The clock was running out, no copilot, it wasn't looking good. I'm thinking, wow, do I let an unknown person come up and help me start a nuclear loaded alert bomber? I wondered briefly if I would catch heat for accepting the suggestion.

Keep in mind, I'm the aircraft commander of the nuclear loaded bomber. I make the decisions. Always leaning forward, hoping to improve our situation I say, "Ground, pilot, send him up."

Up the sergeant comes and gets into the copilot seat. All 8 engine start switches are on the copilot side of the cockpit. He operated the start switches and we both operated the throttles to start the engines. We are now 8 engines running, but still no copilot so still not ready to taxi!

Takeoff in 15 minutes or be destroyed. This was a start engines exercise. After an exercise the Beale command post requested, copied, and tracked various times. There are three possible times: Ready to start engines, ready to taxi, and for a moving exercise, cross runway hold line times. All related to taking off within 15 minutes. During this engine start exercise I needed to respond with our "ready start engines" and "ready to taxi" times.

Shortly after the alert klaxon the Beale AFB command post starts to poll all the Beale and Hill alert aircraft asking for response timing. The Beale and Hill alert aircraft all respond with their times. Each time they polled our sortie at Hill, I responded with a, "Stand By." I mean, I couldn't be ready to start or taxi without a copilot. I didn't want to say that over the radios.

Finally, my copilot crawled up into the plane and hopped into his seat. We were now ready to taxi but well past a 15-minute takeoff goal. I passed our "bad" times to the Beale command post over the radio. We shut down the engines and reconfigured the aircraft back to ground alert status.

We returned to the alert facility and I met with the Lt Col Alert Detachment Commander. I'm a Captain. I told him, "We didn't make the required response timing." He said, "Yes you did. I called the Beale command post on the phone and corrected your times. I saw when your engines were running." (His times meant we made required timing.) I was totally taken aback and surprised by his comment. I said, "We weren't ready to start or taxi because the copilot wasn't onboard. We didn't make the required timing. The times you called in were not correct."

Integrity comes into play here. Our B52 not making required response timing was a big deal. The Detachment Commander, of course, wanted his Satellite Alert command to be successful.

I convened a quick crew meeting. I said, "We can leave the incorrect timing and there is no issue, or, we correct the timing and explain what really happened." My crew agreed with me to tell the truth. I was proud of the crew.

I told the Detachment Commander of our decision, we didn't make the required response timing. The Beale command post was notified of the issue.

The event was analyzed to determine why we missed the required timing. In fact, my copilot procedurally did everything correctly. He was in the Hill AFB movie theater which was on base across the runway. There was a "SAC Alert" light next to movie screen. When the SAC alert light illuminated, he ran out to the alert vehicle and raced to the base of the control tower. There he waited and watched for an expected control tower green light. The green light would give him clearance to dash down the runway and into the alert facility parking area at the end of the runway. He waited but didn't get a green light. It didn't happen. So, he left and drove around the runway perimeter road into the alert area.

Post script. Integrity is important. My crew did the right thing by telling the truth. We didn't make the required timing. As a result of our experience and the analysis, the tower modified their alert checklist to watch for alert SAC vehicles.

We only missed our timing. Had this EAM been a real launch directive we would have been destroyed on the ground. If we had not fixed this, it could have happened to another crew, missed timing or sortie destroyed.

Taking the initiative is important. I didn't get any negative feedback for starting engines with an unknown sergeant, in fact, we were commended for fixing an issue.

I wish I could shake their hands and thank my crew chief, and the engine start sergeant that crawled up into the copilot seat and help me start the B52 engines. Warriors one and all. We were a band of brothers!

CFIC, then the Colonel

Reflecting back in time, in January 1970, I was a 2nd Lieutenant copilot on my first B52 training flight at Castle AFB, CA. I rolled out on final approach to make my first ever B52 landing. I commented over the intercom, "The runway looks too small for the B52 to land on." My Instructor pilot, Major D responded, "Just land your seat on the runway centerline and the plane will land around you on it." I did, and it did. His job as an Instructor Pilot was to keep us alive as I learned to fly it.

Prior to this first flight I had attended weeks of B52 aircraft systems classes and flew many B52 simulator flights. Things happen fast in the Air Force. A few years later on 20 April 1973, I passed my Aircraft Commander (pilot) check ride at Beale AFB. A couple weeks later on 9 May, I flew my 67th combat sortie into Southeast Asia (SEA) from Guam as the aircraft commander with my first combat crew.

Two years later, 25 April 1975, I passed a B52 Instructor Pilot check ride at Beale. The Strategic Air Command wanted all instructors to attend the Central Flight Instructor Course (CFIC) at Carswell AFB, in Fort Worth, Texas.

Completing CFIC was like getting an advanced degree in how to stay alive as an Instructor. I attended CFIC from 27 July through 9 August 1975. The CFIC instructors were amazing. After ground school classes and a simulator, we flew two 5-hour training flights.

CFIC instructors pushed us well past the usual, past our comfort zones, past anything we have ever experienced before. We took off close behind a CFIC KC135 tanker and were soon in contact with the tanker passing 4,000 feet as we climbed. Tanker lead told the departure air traffic control we were a formation of two in close formation. In fact, we were connected with the KC135 tanker via the air refueling boom. After level off we accomplished refueling at extreme bank angles (45 degrees) while climbing and descending. We flew to the triple limit position in the boom envelop. During refueling there are limits on how far the boom can be from center line. Exceed a limit and the boom will disconnect.

While in straight and level flight, the CFIC instructor announced that in 30 seconds he was going to lock the 8 throttles meaning I couldn't move them and adjust thrust. I made smaller and smaller throttle adjustments to maintain position behind the tanker. Then he locked them. I had exactly set the power needed to match the tanker speed and stayed in refueling position.

The CFIC instructors played a student in the left seat, while I was the instructor in the right seat. They did everything wrong, I mean everything. If a student had done it or could have done it while flying the aircraft, they did it to test us. They dropped their checklist and leaning over to pick it up and put the plane into an unusual attitude. They changed aircraft configuration inappropriately. I learned to watch a student at all times and to expect the unexpected. I learned to expect the student to make mistakes but to know when to intervene before anyone got hurt or equipment damaged. To say I learned a lot is an understatement.

CFIC complete, I returned to Beale. Five days later, on 13 August, I was on a routine crew training flight and we were descending into Beale for a full stop landing. We get a call from the command post, "The Wing Commander wants you to pick up a Colonel and give him a few landings."
At this point, I had never actually conducted any training as an instructor pilot. My CFIC training was soon to be tested.

We land and the unknown Colonel climbs into the pilot seat. I don't know anything about him. We chat briefly as we taxied out to the runway. Turns out he is a fighter pilot who has never been in a B52 before!

I put his hand on the eight throttles, my hand was on top of his, and we push the throttles up to takeoff setting. We have lift off and are soon on downwind. I'm talking power, airspeed, and trim setting to him. We roll out onto final. I'm talking about the big trim changes required during the landing flare. As we neared the runway, I'm calling over the intercom, "Trim, trim, trim, trim!" His trim changes weren't large enough. At the last moment from touchdown I grabbed the control column, and in a desperation act, pulled it way back toward my chest.

The plane landed and I accomplished the touch and go checklist: Airbrakes 6, stab trim reset to takeoff setting, airbrakes off, standup the throttles, power checks, and takeoff power set.

We did this a second time with my, "Trim, trim, trim" calls on short final and another desperation pull back on the control column for landing flare. Finally, we did a full stop landing. After parking and after I shutdown the engines, the fighter pilot Colonel crawled out of the pilot seat and patted me on my left shoulder saying with a nice smile, "Thanks."

The fighter pilot Colonel was gone, the plane was shut down, I crossed my arms over the control column and laid my helmeted head down on them thinking, "I'm still alive." I credit my CFIC training with keeping me alive, barely.

Beale ORI

In summer of 1975, the Beale Bomb Wing had been notified that in the summer of 1976 it would stand down. The B52 Wing would leave and the U2 Wing from Davis-Monthan would move in. The B52s and crewmembers would be disbursed to other Bomb Wings located all around the country.

SAC exempted us from all aircrew training requirements (ground and flight) and command evaluations. We used this unusual situation by conducting upgrade training of all positions. I was an instructor pilot. I put my copilot in the pilot seat and gave him pilot upgrade training on every flight. Navigators received training to be a radar navigator. All positions trained to become instructors. When our crewmembers were dispersed to other bases, they would arrive with a stack of upgrade training documents. This would, hopefully, move them ahead of their new unit personnel.

I flew a number of pilot upgrade flights with various squadron copilots. As an instructor, I let trainees make mistakes that they learn from. The important part was to know how long to let the mistake continue until intervening before equipment was damaged, or people hurt.

I was in the copilot seat doing upgrade training with a copilot in the pilot seat. We were on downwind for a visual touch and go. I decide to see if he is paying attention to the checklist so without actually putting the landing gear down, I called "Landing gear down, six down and checked." There is a landing gear lever that is used to put the gear up or down. Next to the handle are 6 indicators that display three gear possibilities: Up, in transit, or down. He is supposed to look and check them down also, then respond, "Six checked and down." He didn't check them but responded, "Six checked and down."

We turn base leg. I call for another gear check, and again, without looking he responds that they are down. We turn onto the final approach and we are heading for the runway. I call for another gear check and once again he doesn't check. We are heading toward the runway and the landing gear are not down. It was time to intervene. I tugged on his shoulder and pointed to the gear indicators. His eyes got big! I gave him a thumbs up and then lowered the landing gear. He then correctly checks them and made his gear down call. I'm confident he would never make that mistake again.

On another day I was giving my copilot a left seat upgrade training flight. He was doing great throughout the preflight, engine start, and taxi to the runway hold line. At this point, he missed an important item. There is a number three window next to both the pilot and copilot positions. They are triangular in shape. Its function is to be able to open on the ground so the cockpit pressure is the same as the outside pressure. It would be unlocked, pulled inward, and slid aft. The nav team always called and checked that it was open before they unlatched the crew entry hatch. If the cockpit was pressurized and the entry hatch unlocked, it could blow open and injure someone.

My number 3 window was locked into place ready for takeoff. His wasn't, it was unlocked and positioned inward toward him. It was time to learn a lesson. Pre-takeoff checklists complete, I asked him if he was ready for takeoff. "Yes," he said. "Are you sure?" I asked. "Yes," he responds. He added power to taxi out onto the runway and aligned to the centerline. As he pushed up the throttles to takeoff power setting, to his surprise, his number 3 window slid aft! I said, "I have the plane" and continued the takeoff. He struggled to pull the window forward and put it into the locked position. He locked the window in place on the runway and took the plane back. Lesson learned.

With respect to the exempted evaluations. Our Bomb Wing Commander volunteered the Wing for a SAC Operational Readiness Inspection (ORI)! This was an eye opener. It was like volunteering to take the biggest test that ever existed.

The Inspector General (IG) team arrived. The evaluation started with a recall of the all Wing personnel and all the B52s were configured to nuclear war status. We demonstrated our ability to generate all aircraft according to the published plan. The Generation Phase over, the nuclear weapons were downloaded. The aircraft and crews prepared to fly. The crews went into crew rest.

The 13 aircraft stream would takeoff in 15-minute intervals. We were scheduled number three. We had completed the aircraft preflight and sat strapped in our ejection seats waiting for our start engines time. A car stopped in front of the plane. An IG member crawled up into the plane. He strapped into the Instructor Pilot seat between and just behind the two pilot's seats. He came up on the intercom, "I'm going ride with you on the flight as an observer." I gave him a smile and said, "Welcome aboard." I'm actually thinking, I don't know anything about this man and was not happy having him on board.

We start engines and taxi out toward the runway. The IG person and I are talking about aircraft. I mention on how much more reliable the B52G systems were compared to the B52D. We are cleared for takeoff. Ironically, when I set the throttles up to takeoff power the Master Caution Light immediately illuminated indicating a system problem. I aborted the takeoff (didn't deploy the drag chute), turned off the runway, and taxied into the parking ramp.

We sat on the ramp with the 8 engines running while maintenance fixed the problem. It was a minor generator problem. Maintenance reset a circuit breaker in a wheel well that fixed the issue. We were taking off into the peace time world, I aborted, if this happened in the cold war world, I would have taken off.

We would now takeoff last in the stream, behind the remaining ten B52s. We would be sitting on the ramp for 2 ½ hours! I called the command post, "Control, do you want me to shut down engines?" The command post responded, "No, the Wing Commander wants you to leave your engines running." This was the right response. I was now in an engine running aircraft with all systems running. Shutting down obviously meant that I would have had to restart the entire aircraft again with more possible start problems.

We sat in silence awhile. Finally, the guy in the IP seat asked, "How will this impact the flight?" My copilot and I looked into each other eyes. I realized this was the opportunity to get rid of him. I fibbed to him saying, "Well, because of our ground delay and fuel burn, we would probably need to divert and land at an east coast base."

Shortly he says, "I need to leave." We shook hands and I said, "Sorry, I would have liked to show you a good ride!" He crawled down out of the plane. Yes!

We took off last, flew the last sortie in the bomber stream, and landed last. Ironically, in addition to four gravity bombs, we carried an AGM28 Hound Dog missile under each wing. The two missiles needed to be programmed. An hour programing leg was flown where the navigator team computed and inserted data into the missiles guidance system. Once the guidance system was programed, in theory, it should fly the missile to its targets.

If we had taken off number three, as scheduled, we would have flown the missile programing leg at the worst possible time, sunset. Because of our delayed takeoff we programed the missiles later at night. We had the best Wing AGM28 missile scores!

Little America

Through the years I upgraded to Instructor Pilot, and then on 22 September 1975, was selected for Standardization/Evaluation Division. This time as a Bomb Wing evaluator (Check Pilot) on crew SO3. We gave squadron crewmembers check rides, flew training flights, and sat nuclear alert.

Want to ride along on a B52 training sortie? Have you ever been airsick?

This description is for those not familiar with the pressurized cockpit layout of the B52. Back then, only the two pilots had windows and could see directly out of the B52. About 20 feet to the rear were the aft facing electronic warfare officer and gunner stations. They would eject, if needed, upward. Just behind them was a square opening and ladder down to the lower deck. A few feet forward of the ladder was the forward-facing radar navigator and navigator stations. Their ejections seats ejected downward.

There were low level training routes located all around the states. At the end of the routes were Radar Bomb Scoring (RBS) sites to score simulated bomb releases. We didn't actually drop anything. Rather, we turned a UHF radio tone on beginning 20 seconds before simulated release. The RBS site locked onto the aircraft and monitored the tone. At weapon release the tone stopped. Weapon accuracy was calculated from the stop tone point. The routes gave SAC crews opportunity to train low level tactics needed to attack the Soviet Union. Occasionally, at least yearly, a new route would open. Around 1975/1976, Little America low level route and bomb plot, became active. My crew was selected to be the first Beale crew to fly it.

It was a typical 8-hour training sortie profile. We accomplished a celestial navigation training leg. The EWO pushed a sextant up through the top of the plane and "shot" the sun in a set 3-minute sequence. He passed the data over the intercom down to the navigator who used the data to calculate our position.

After the navigation leg we rendezvoused with a KC135 and accomplished about 30 minutes of air refueling practice.

Next came the new "Little America" low level route. The purpose of flying low level was flying below enemy radar where they can't paint (see) the B52 on their radar. High terrain masked (hides) the plane from being seen. Also, ground radar functions are line-of-sight so even over flat terrain, depending on the distance from the transmitter, the plane may be below the radar. This is because of the curvature of the earth.

As you read this remember that my crew were experienced veterans with lots of hours in the B52.

We descended into the low-level route. It was day and visibility were great. I'm flying Terrain Avoidance (TA) using an aircraft radar generated terrain trace on a scope on the instrument panel to keep us above the ground at 400 feet. It was a very turbulent rough ride. Crosswinds across ridges cause turbulence. As I pulled the nose up to climb up the terrain, we got heavy as it caused a positive G-force pulling us down into the ejection seats. As we crested over the ridges at 400 feet, I pushed the nose down seeking to return to 400 feet above the ground. This resulted in a negative G-load and we grew light in the seats. Had we not been strapped in we would have floated off the seats.

For those who have never been on a flight like this just imagine you're in a closet (can't see out) that is strapped onto a roller coaster, except this ride is more violent.

I have the easiest position in the B52. I can see what is happening, what's going to happen, and know when I'm going to accomplish positive/negative maneuvers. The others only feel the effects as they occur. About halfway on the turbulent route the aft facing electronic warfare officer and the gunner became airsick.

As we approached the bomb run, I note that only one voice is reading the "Weapons Release Checklist." The checklist is supposed to be challenge and reply. I should be hearing two voices but I only hear the Navigator:

"Bomb Inhibit Switch?"	"Off"
"Special Weapons Lock Indicators?"	"Indicated Locked"
"Release Circuits Disconnected."	"Connected, Light On"
"Bomb Release Indicator Circuit."	"Select, Light Dim"
"Master Bomb Control Switch?"	"On, Light ON"

I ask over the intercom, "Nav, pilot, what's up?" The navigator responded with, "The RN has his head in his helmet bag." In other words, he was airsick and vomiting in his helmet bag. I asked, "Nav, you got this?" He answers, "Yes." He and I continue the bomb run.

At that point, half of the crew was not functional and I'm sure very miserable. I'm thinking that this is just a training sortie so maybe I should abort and climb out, but then on the other hand, this could happen on a real-world nuclear war sortie, so I continue.

We were a just few minutes from bombs away. I'm hunkered down in my seat with eyes on the gauges concentrating on the maintaining correct airspeed, the TA terrain trace (keeping us 400 feet above the ground), keeping the FCI centered (guidance to the target), and monitoring the Time-to-Go (to release) counter. A low-level TA bomb run is an intense event!

I get a couple of tugs on my right arm sleeve and I glance quickly over at the copilot. His helmet framed his face. The oxygen mask was hanging off the left side. Eyes were big and he had both glove covered hands are over his mouth. There was vomit oozing, squirting, out between his fingers. Obviously, he couldn't talk on the intercom or function as a copilot. Now it was just the two of us, pilot and navigator functioning.

I go back to concentrating on the flight instruments. I gave the copilot a fist thumb pump toward the rear meaning to unstrap and get out of his ejection seat and move back. I didn't look over again during the remainder of the bomb run.

I had two copilots assigned to my crew. The second was sitting in the instructor pilot seat just behind and between the two pilot's ejection seats. The copilots did a seat change as the bombs were released!

Timely Decisions

A month later, Friday, 10 October, I walk into the Standardization/Evaluation office late in the afternoon. I find Harold hunched over his desk intensely reading something. "Harold," I ask, "What are you working on?" He says, "I'm going to drive down to Sacramento tonight to attend a Pepperdine University Master's Degree program and I'm reading up about it." I say, "Tell me about it." He says, "It's a 10-month program. Pepperdine will fly their professors up to Sacramento. Every other week classes are held at the Sacramento Inn on Friday nights, and Saturdays. Students will be military members from the Sacramento area. The degree is in Sociology."

I'm thinking 10 months! The Bomb Wing was scheduled to close in about 10 months, summer of 1976. All the crews and B52s would be dispersed to other units. We were leaving and the U2 operation was moving in. There was barely enough time to complete this before having to move. I could do this.

One way to enhance a promotion potential was to complete a Master's Degree. In reality, I was competing with all the other officers in my age group. Be promoted and stay in the Air Force longer. Get passed over and separate sooner.

My Bachelor of Science degree was in Sociology so I'm familiar with the topic. I said, "Harold, I can't attend tonight, but tell them I'll be there tomorrow morning." This conversation took about 3 minutes. See an opportunity and take it!

Ten months later the Bomb Wing stood down. We were able to delay leaving Beale a few weeks until the Master's Degree program was complete. Pepperdine sent us outstanding professors. On 17 July 1976, I underwent Orals. It reminded me of my survival school prisoner of war interrogation. Then on the 23/24th we wrote comprehensive exams. I wrote 42 pages of answers!

Rule one, always screw with the "Friggen" New Guy, the FNG.

We lived in a live and/or die world, yet there was time for humor. Most was related to the new personnel. The Friggen New Guy, the FNG.

I was in the Strategic Air Command B52s and KC135s for 16 years. I entered the Air Force in 1969. In 1992 the Air Force was reorganized and SAC stood down. The Cold War nuclear alert was serious business. During the cold war deterrence was key that prevented nuclear war with the Soviet Union. As a B52 crewmember, I spent, on the average, 10 days a month on nuclear alert with a requirement to be able to takeoff in 15 minutes! Acknowledging the seriousness of alert, there was still opportunities for mischief.

The white eye patch!

Pilots had a number of ways to avoid being flash blinded by a nuclear bomb detonation. There were thermal curtains that completely covered all the pilot's windows. The curtains were pulled closed just after takeoff. Each pilot/copilot thermal curtain had a small flap that could be pulled open to peek out ahead. There were divider curtains so the pilot could not see out of the copilot opening and the copilot could not see out the pilot's small opening.

To look out the small opening we had a hard cover white eye patch that would cover one eye so we weren't totally blinded when looking out the small opening as a nuclear detonation flash occurred. Only one eye was blinded. We had four eyes in the cockpit.

The FNG attended flash blindness training and learns about the eye patch. We tell him that for training purposes he needed to wear the white eye patch all day. Well, that same afternoon he had an appointment to interview with the Bomb Wing commander. In he goes with the white eye patch on. The wing commander has seen this joke played on many others before and says, with a knowing smile, "So I see you had flash blindness training today!"

The Waiver

The war plan against the Soviet Union was the Single Integrated Operational Plan (SIOP), Emergency War Order (EWO). All SAC units were assigned sorties covering specific targets. The last event for the FNG to become combat ready was to "Certify the SIOP" by studying and briefing a specific unit assigned EWO sortie. Sortie study lasted many days. Each classified sortie bag contained specific sortie information. Charts and maps of the routing, both high altitude and low level, and air refueling locations. In addition, there was bomb run information headings and timing, and target study. Finally, there were post-strike charts and recovery information.

Usually a 6-man crew would certify together with each member briefing on his specific position. The aircraft commander would start the briefing, then the copilot, navigator, radar navigator, electronic warfare officer, and finally the gunner. Any, or all could be an FNG. A Colonel (Wing Commander, Vice Commander, or Deputy Commander for Operations) received the briefing. There were usually 20-25 wing staff members from the various functional specialties seated behind the Colonel. When the certifying crew finished briefing, the Colonel started asking questions. Then questions came from the staff. The Q and A went on for quite awhile.

At some point the Colonel was satisfied they knew and understood the sortie. He signed certification certificates for each crewmember.

The FNG then signed a small pile of official documents. He eventually came to a "Waiver of Cannibalism" document. It said he agreed that in a survival situation, the other crewmembers could eat his body. There was a list with suggested recipe titles:

Butt Cheek Steaks
Rump Roast
Finger Licking Good Fingers
Toe Knuckle Tidbits
Tender Ribs and Gravy
Miscellaneous, Mystery Stew

Most of the FNG's stared at the document in disbelief but then signed it!

The water dump!

I had an FNG copilot on his first nuclear alert tour. The B52G used demineralized water augmentation to increase thrust for takeoff. Water was literally squirted into the engines during takeoff. In the B52G, the water tank was just forward of the wings, just behind of the pressurized crew compartment.

When the temperature decreased to about 28 F degrees and was forecast to be there for more than 3 hours, we needed to dump the water to prevent it from freezing. There was a single dump port located high on the left side of the fuselage. The dump port was about 2 inches in diameter. The water came out in a solid massive stream.

One night we get the call from the Command Post to dump water. We take the FNG copilot out and station him next to the aircraft where the water stream will hit him. We tell him to stand there and watch the number three engine. Then we go up into the cockpit and turn on the battery and hit the dump switch. The copilot just got hammered with the ice-cold water stream! Welcome to the crew.

The Echo Check!

B52 gunners were a special breed. They were willing to do anything to help the mission. One week I had an FNG gunner on his first alert tour with my crew. SAC units had the weekly alert exercise. The Klaxon was a horn that blew. Alert crew response to a Klaxon was to respond to the aircraft and started engines. We lived in the Cold War nuclear war world. During an alert response the crew doesn't know if the Klaxon was an exercise or real-world threat until we got into the aircraft, copied, and decoded the Emergency Action Message.

This Klaxon was an engine start exercise. The aircraft now needed to be re-cocked back to alert status. Engines were shut down and new start cartridges were loaded into the engines.

As this was occurring my eager to help, FNG gunner departed the cockpit and came up on the crew chief headset. This was unusual and not the gunner's re-cock responsibility, the crew chief owned the headset. The gunner was down to my left on the ground looking up at me. He asked, "What can I do to help the crew chief re-cock the aircraft?

Well, there was an engine check on the re-cock that was accomplished. It was a functional test of the two engine igniters in each engine. The crew chief would stand in front of each engine and listen for the two-engine start igniter's popping as we turned each engine starter ON. The gunner knew something needed to be done at the engines during re-cock. I applied rule one, screw with him.

I told the gunner, "Well, you can do the engine echo checks for the crew chief." He says, "What do I need to do?" I tell him, "Get in front of number 1 engine," and he did. I told him, "Now yell as loud as you can into the engine and listen for an echo. If you hear an echo then we would need an engine stabulator (no such thing) changed out."

He yells and says he didn't hear an echo on number one. Moving over to number two engine he yells again. No echo there either. By now the crew chiefs and security guards are knowingly grinning as they watched the gunner. Engines 3 and 4 are higher off the ground. I told him, "You'll need a ladder to check engines 3 and 4 so ask the crew chief for one." He does and the crew chief asked, "Why?" The gunner responds, "To complete the engine echo checks." The crew chief pointed at a ladder over at the power cart. My FNG gunner sets up the ladder in front of engine number three and up he climbs.

By now everyone on the ramp is laughing and the gunner looks around and realizes he had been had! He looks up at me and gives me the finger! At that moment he became a SAC crewmember on my crew. Gunners are a special people.

Who buys the beer? The FNG of course.

During the war in South East Asia I flew both B52Ds and Gs out of Guam and Utapao, Thailand. During the crew aircraft preflight, chalk lines were drawn on a forward landing gear tire creating 5 sections that were marked: P, CP, RN, N, EWO. The deal was, when the plane stopped and was parked whatever crew position was on the ground bought the beer that night. The FNG fix was in, prior to flight I told the crew chief which position needed to be on the ground when we parked. This worked for a few flights until the FNG figured it out.

Our Squadron Losses

My assignment to Beale was outstanding! I worked hard, survived, and thrived. However, there were trying times when we experienced loses. We all did the same things but some didn't survive. Beale loses were typical to what other B52 units experienced all across the country. During my 6 years at Beale, 18 fellow squadron crewmembers died. Think you have had a bad day? Here is what can happen to aircrew members.

During the South East Asia war five died flying a combat sortie out of Utapao AB, Thailand. Only the gunner survived. Cell leader led them into a thunderstorm over northern Thailand. At 32,000 feet, they lost flight instrument lighting, then aircraft control.

There apparently was a series of survival equipment failures: The navigator ejected downward, however, he struck the lower fuselage and then he didn't separate from his ejection seat. He rode the seat to the ground without a chute opening. The radar navigator was not strapped into his ejection seat, he rode the aircraft to the ground. The electronic warfare officer ejected but was hit in the head by the pilot's ejection seat. The pilot ejected but died from flail injuries after separating from his seat. The copilot ejected but his right parachute riser Q1 quick release was open at ejection meaning that it released from his chest and the chute streamed above him as he dropped to the ground.

The gunner in the tail survived! Electrics was lost so there was no communication with the crew up front. He didn't know what was happening! A red bailout light illuminated, then he saw ejection hatches, seats, and crewmembers fly past him. The gunners escaped by jettisoning the tail end of the plane. It included the 4-50 caliber machine guns, ammunition, and gun radar. An oval hole appeared just in front of the gunner and he physically jumped out of it. He then manually deployed the parachute.

As the gunner descended, storm air thermals took him up and down a few cycles. Finally, he came out of the bottom of the thunderstorm and landed in a Thailand rice field. Thais approached him on the ground, one wanted him to go somewhere. He stood his ground and pulled out his .38 6-shot revolver. He talked on his handheld survival radio as if someone was listening, at first nobody was. Thais started to take his survival equipment that was spread around him on the ground. Finally, a black helicopter came up on his frequency and a few minutes later picked him up. That was his last ever B52 flight.

Four crewmembers died when their B52G was shot down by a Surface to Air Missiles (SAMs) over Hanoi in December 1972. The two survivors were captured and became Prisoners of War.

Nine died during a taxi back takeoff at Beale AFB, CA. The crew completed an 8-hour training sortie and landed at Beale. The pilots needed to complete their annual check rides. Evaluators were picked up to complete the checks. I'm told that on takeoff, fuel pump malfunctions in a main tank resulted in the four right engines flaming out. The thrust from the left engines shoved the B52G off the runway to the right.

The pilot, crawled out of the wreckage, but became covered with JP-4 jet fuel. It ignited. He was burned everywhere but his helmet protected head. He lived only a couple days. A gunner got out but died as he ran under a wing. An explosion dropped the wing on him. The other seven crewmembers were found in the area of the crew entry hatch.

The Strategic Air Command usually lost 1 B52 and crew a year during routine training.

During my 6 years at Beale (1970-76) I progressed from being a brand-new copilot, to instructor/evaluator copilot, aircraft commander, instructor pilot, and finally evaluator pilot. I flew 78 combat sorties in South East Asia in B52D (Arc Light) and the B52G (Bullet Shot). I dropped bombs (500 lb., 750 lb., and cluster bombs) in South Vietnam, Cambodia, and Laos. I dropped just under 5,000 bombs on Nouns.... that is: a person, place, or thing.

The Beale Bomb Wing closed and aircraft and crewmembers were dispersed to other SAC units. On 12 August, 1976, I left Beale and drove to Fairchild AFB, Spokane, WA.

About a month later, I received a letter from Pepperdine University saying I passed the Master's Degree program. I have heard that not everyone passed!

Chapter 4 – Career Broadening

Command Post

From Beale, I was assigned to the 325th Bomb Squadron at Fairchild AFB, Spokane, WA. The former Beale Wing Commander had been moved to Fairchild. I believe he had arranged my assignment there.

When I arrived at Fairchild, I hadn't flown a B52 for 4 months. I think I intimidated the 325th Bomb Squadron senior instructor pilot when we flew my first training sortie at Fairchild. I'm in the pilots (left) seat and he is in the right seat. I complete the standard air refueling off load training. I then tell the boom operator, "I want to fly a triple boom limit." He responds, "Ok."

To refuel you have to keep the aircraft within a boom envelope (a box). There are right/left, up/down, and in/out limits. Exceed a limit and the boom automatically disconnects. The pilot or boom operator can also manually disconnect.

I slide over to a right limit, then down to the lower limit, and finally moved aft to the aft limit. I'm at three limits of the air refueling box. The tanker boom operator voice over the radio expressed total amazement exclaiming, "You did it!" I responded, "Recovering." I returned back to the middle of the envelope. A triple boom limit wasn't typically needed, flown, nor required. It wasn't something that was usual or easily accomplished.

The instructor pilot in the right seat asked, "Can you do that from this seat?" I respond, "Yes, I'm an instructor pilot. I can do it from either seat." And then I asked him, "Do you want to try a triple limit?" He answers, "No." I actually wondered if he had attended CFIC.

After the flight, I wonder what he said to the 325th Bomb Squadron Commander about me because a couple days later, the Bomb Wing Deputy Commander for Operations (DO) called me into his office. He said, "I want to move you from the Bomb Squadron to be a Bomb Wing command post officer controller." We discussed the offer a bit then I told him, "I'll serve wherever you need me."

At that point, I had been in the B52 flying world for 6 years. I'd been to war, an instructor pilot, and a stand/board check pilot. I had served in all possible squadron positions. I was extremely proficient, at the top of the B52 flying game.

My perception at the time was that being assigned to the command post was not a good career move. Through the years I had communicated with the command post on the UHF radios every time I flew and while on nuclear alert. The command post would be a totally new world for me. Ultimately, this command post experience (Career Broadening) proved extremely valuable to me in the future years.

The next day I was walking along a sidewalk on base and the Wing Commander (my former Wing Commander from Beale) pulled over and stopped. He invited me into his car and asked, "Are you sure you want to go to the command post?" I responded, "That's where the DO wants me so I'll make the move." The DO worked for the Wing Commander. I'm confident that if I had told him that I didn't want the move, he would have cancelled it.

I served three years in the Fairchild command post. The position was multifaceted. We were the Bomb Wing coordination center for almost everything that happened on the Base and in the Wing. As issues occurred, we determined how to respond. We had many Quick Reaction

Checklists (QRC) covering various issues to guide and help us. When an event occurred and we usually had a checklist to accomplish. Many events required coordination with SAC Headquarters via phone or messages.

Our first priority was supporting the Single Integrated Operational Plan, the Nuclear War. I'm talking cold war nuclear deterrence. We assisted the nuclear alert crews to maintain their aircraft on alert and were a conduit for the relaying Emergency Action Messages over the UHF radios to the alert crews.

We also worked issues with routine aircrew training flights. Having been a B52 instructor pilot, I knew the B52 inside and out. Crews would call in with a maintenance issue over the UHF radio during preflight or inflight. I'd relay the issue on a landline to maintenance Job Control. They would respond with the required maintenance specialties.

Command Post Controller Stations

I called the senior staff (Wing Commander, Deputy Commander for Operations, Deputy Commander for Maintenance, and Base Commander) day or night with all sorts of events. I learned quickly to call them with a suggested solution for approval, not just with the problem. I would coordinate with various wing agencies to determine feasible courses of action. I then would call with the issue, options, and my recommendation. They would usually concur. They trusted me. (Years later I was a Deputy Commander for Operations and received command post calls day and night.)

Initially, I worked on the controller console as an

Officer controller. There were also two airmen on duty, one on the control console as enlisted controller, and one who worked and input electronic messages and reports to Headquarters. These young enlisted men were awesome.

The officer and airman consoles were identical. We each had a handset phone and a panel with square hot buttons to all the senior staff and all agencies on base. There was a red phone between us that was a direct hotline to the Strategic Air Command underground command post in Omaha, NE, with the SAC Airborne Command Post, and with 15th Air Force, CA. We also had buttons to talk with aircraft on two radio frequencies.

The command post was a restricted area. We wore .38 caliber 6-shot revolvers and controlled the command post entry door. There was top secret nuclear war execution material in the command post.

When someone buzzed from the hall outside the door, there was a handset and a one-way window so we could both see the hallway and talk to the person at the door. We could buzz the door open to allow entry.

There were four rooms in the command post: Our elevated controller console, two administrator rooms, and a Wing Battle Staff room. Our elevated console looked down through a glass wall into the command post administration room.

Along the side of our console area was a nook that contained a safe. It contained highly classified nuclear war materials. We didn't have the combination. The wing organization that used the material needed to access the safe through us. This arrangement was part of the security in place. Anyone who knows me, knows I have a rule in my life. Always screw with people when you can, that's my rule. One day an officer arrived to conduct an audit of the top-secret contents in the safe. He went into the nook next to us and pulled a curtain closed to prevent us from seeing what he was doing.

There was a secure phone sitting on top of the safe. I decide to screw with him. I called the secure phone number from my controller console just 20 feet from him. He answered the secure phone. I said, "This is the SAC Command Post with a test of the secure communications check. Please count down from 50 to 30." He is counting down, "50, 49, 48, 47." I then interrupt him saying, "Now count up from 10 to 20." He is now counting up, "10, 11,

12, 13." I then ask him, "Now count down again from 30." At this point he pulls the curtain open, looks around and sees me! He knows he had been screwed with, he smiles, and then gives me the finger. (Years later I had access to the material in the safe. During that time, I was prohibited from traveling outside of the US without specific approval.)

There was a wall of shelves in the command post that held Operations Orders and Plans that the Wing was committed to support. When I first arrived, I reviewed all of them. I recall one, the 5027 Plan. It was the defense of South Korea.

We worked on a three 8-hour shift schedule. When someone was away on leave it was a 12-hour shift. It was, at times, brutal. The first two days were two swing shifts 2:45 pm to 10:45 pm. Then two-night shifts 10:45 pm to 6:45 am. Finally, two-day shift days 6:45 am to 2:45 pm. This cycle was followed by a couple days off.

The 47th Air Division was collocated at Fairchild. The 47th Commander was a general who occasionally flew to Offutt, AFB, to serve as the Airborne Emergency Action Officer (AEAO). A SAC Airborne Command Post (SAC ABNCP) had been airborne since 1961. They flew 8-hour shifts. It was a Cold War backup for the underground command post at Offutt, AFB. If the Soviet Union destroyed the underground, the AEAO with the airborne command post could direct the nuclear response. It was a battle wagon! I accompanied him to Omaha and flew on an orientation 8-hour flight. (Years later, I helped put the SAC Airborne Command Post onto Ground Alert)

After the orientation flight, I toured the SAC underground command post at Offutt AFB, Omaha. There were direct hotlines to all the command posts around the world.

I quickly seized another opportunity to screw with others. My sense of humor was immediately switched on. I had the Fairchild command post connected on a direct hotline and I announce to Fairchild, "This is the SAC Command Post conducting a physical fitness test of all officer controllers. All officer controllers will accomplish maximum pushups in one minute. Standby to report." It didn't work. They recognized my voice and just laughed.

The visit to SAC was truly informative, I learned a lot about how the command and control network worked.

When I returned to Fairchild, I briefed the other controllers about my observations.

Fairchild AFB had a unique fog dispersal system on the approach to runway 23. I have never seen one, nor heard of another one, anywhere else.

Instrument approaches guided pilots down to 200 feet above the ground. For a pilot to land the aircraft, the pilot must see the runway environment at 200 feet. If at 200 feet the runway wasn't seen, then a missed approach was made.

The fog dispersal system consisted of an array of propane tanks (10-12) spread out from the runway along the ground track of the instrument approach.

Each tank had a 10-foot pole. Propane was released from the top. The system only worked in specific weather conditions. It would be activated in freezing weather in dense fog. Propane gas molecules attracted fog water molecules, they then froze into ice crystals, and dropped to the ground. We called it snog.

Inflight visibility improves, but the ground became icy. In other words, the result was that Runway Visibility Reading (RVR) increased, however, the Runway Condition Reading (RCR) decreased as it became more slippery. Airway Heights was a small community along a highway. It was near the approach ground track. I wonder how many vehicle accidents occurred there as a result of the icing of the highway.

We occasionally had personnel from organizations across the Wing come into the command post for a command post orientation. On these visits, there were at least 6 folks standing behind me as I explained what we did in the command post. Our communication phone panels had about 40 square push hot buttons with an organization label under them.

To screw with them I put a tape under a nonfunctional red square hot button that said, "Building Self-Destruct." As I briefed the folks and pointed to the various communication panel buttons, I would pass over the self-destruct button without explanation. They usually stopped listening to me at that point and just starred at the self-destruct button.

One of the other command post officers was a true careerist. He always ran scared, career worried. When something bad happened in the wing he had the airman controller make the wing notifications to the senior staff. When something good happened, he made the notifications.

He was the officer controller on duty when a unit practice Operational Readiness Inspection began. The entire Wing was recalled to the base and all aircraft were generated to alert status.

The command post personnel gathered in the administration office below the controller console. It was time for me to screw with the careerist. A KC135 crew arrived at the command post door. They had a legitimate reason to enter the command post. He buzzed the door open and let them in. I wrote a note, folded it in half, and gave it to one of the tanker crewmembers asking him to hand it to the officer controller.

My note read, "Make no notifications, this is an exercise, the command post has been infiltrated and all personnel killed. Make no notifications."

All the command post personnel in the admin room below him were watching his reaction. The look on his face was just awesome, total shock, eyes big, mouth hanging open, he knew his career was over. Then he looked down through the glass window at us laughing in the admin room and he knew he had been screwed with.

On 1 January, 1978, I became the command post Training Officer. Within the command post, I trained the new personnel and conducted a monthly command control recurrent training. I also taught initial Command Control Procedures training to new B52/KC135 crewmembers. (BTW, I trained the first women KC135 pilot assigned to Fairchild.)

I had a desk in the command post side admin office where I produced my command and control training products. My desk butted up against the command post bosses' desk, we sat face to face. He smoked a pipe. I've never smoked a thing in my life. His smoking was disgusting. Tobacco and ash would drop on his shirt front. Occasionally sparks would burn holes in his shirt. He, of course, filled the room with smoke. It was tough to sit there.

I had about a year and a half remaining on my command post tour. After 6 months I decided to act and, you guessed it, screw with him. I began to put tabasco sauce on his pipe mouth piece. He continued to smoke however he kept grumbling and complaining about how bitter and nasty tasting pipe smoking was. He never figured it out the remaining year, that is, until my farewell party.

Every seven days the nuclear alert force B52 bomber and KC135 tanker crews changed over. I went to the alert facility where I conducted refresher training with the oncoming crews and reviewed nuclear command control messages and procedures. In preparation, I recorded a series of 5 Emergency Action Messages (EAMs) on a tape recorder followed by a few questions asking the crew response was to the EAM. I played the tape and they copied the messages and answered the questions. I then explained each message and the correct response.

About two years into my 3-year command post assignment I contemplated my future. More life decisions to make. Should I stay in the Air Force or leave for the civilian world. One possible option was to apply to the airlines.

Ultimately, three times in my life I made the decision not to apply to be an airline pilot. Life career decisions are always ongoing. After graduating from flight school, I incurred a four-year obligation to remain in the AF. I completed the obligation and was now free to leave active duty if I desired. The airlines were hiring. Air Force pilot retention was directly impacted by the airline hiring.

In 1976, I was 29 years old and the airlines were hiring. Back then, the airlines viewed anyone 30 years old as too old to hire! It was decision time! I decided no. With the no decision I believed that I would never ever have an opportunity to apply to be an airline pilot.

Back then, Air Force pilots had to be careful. If it became known you were interviewing with airlines, your Officer Effectiveness Report (OER) could be impacted. The AF leadership rewarded career members. If not hired by an airline, then a negative OER would impact their AF career. Pilots would disappear for a few days to attend airline interviews. I saw Fairchild pilots literally sneaking through the Spokane airport hoping they didn't run into someone they knew.

Believe it or not, I had two more opportunities to fly in the airlines. Later in life, airline pilot hiring situation changed. Age was no longer an issue. At my 20-year point (age 42) I could have retired and interviewed for the airlines. I once again elected to remain in the AF. Even at age 52 when I hit the 30-year mandatory retire point I could have been hired.

I have never regretted my decisions. I don't think I

would have enjoyed an airline pilots' life.

About two years into the three-year assignment I was on the command post console. It was a quiet day so I went next assignment hunting.

Thinking back now, after graduating from college, I don't know how or why I was assigned to attend pilot training at Randolph, AFB, TX, there were many flight school bases. Then after flight school graduation how I was assigned to B52s at Beale, AFB, there were many B52 bases around the country. Then, from Beale, I think the previous Beale Wing Commander had me assigned to Fairchild AFB. There was an assignment input opportunity for everyone. On the Form 90, we could list three bases and jobs that we desired. We called it the Dream Sheet.

I had learned to be proactive so I called around to the various Air Force assignment shops looking for my next assignment.

I was in the Strategic Air Command so I called SAC assignments first. The voice suggested moving from a unit command post to a SAC Numbered Air Force or SAC Headquarters Command Post. Or, the SAC voice said, "How about a B52 instructor pilot at Castle, AFB, Merced, CA?" This made sense from their perspective but wasn't from mine. It didn't show progression.

I then called the Exchange Program assignment folks. The voice said, "Sorry, but foreign countries wanted fighter pilots." Next, I called Air Force assignments, the response heard was, "We have nothing for you." Things were not looking good.

Finally, I called the Department of Defense/Joint Assignments (DOD/JT). Staff Sargent Browning answered. I said, "I don't even know what a DOD/JT assignment is but do you have anything for me?" He asked, "What is your Social Security number?"

A few minutes later he came back online and said, "I have these 5 assignments you can select from. If you were a Major there are many others." He read off the five:

US Embassy, C12 pilot in Tehran, Iran.
US Embassy, C12 pilot in Athens, Greece.
United Nations Truce Supervision Organization (UNTSO) (Middle East)
Commander in Chief, Pacific, Airborne Command Post.
SHAPE Headquarters, Mons Belgium.

Totally startled, and in a slight daze, with his response I asked, "Did you just say I could pick one!" Sargent Browning said, "Yes, I make the DOD/JT assignments and you can have your pick." I asked a couple questions about the assignments and then asked for a couple of names and phone numbers of officers that served in the assignments. Finally, I asked, "Can I have a couple days to answer?" "Yes," he responds.

I made a few phone calls asking about these assignments. They gave me a good understanding of each. I called Sargent Browning back and said, "I'll take the UNTSO assignment." The UNTSO assignment was just a little over a year away.

A few comments about assignments and promotions: A few years earlier the Air Force changed the Officer Effectiveness Report (OER) (think job performance report) system structure. The older 9-4 system had inflated to the point that everyone was receiving the highest possible rating (9-4). The OER's were reviewed during promotion board considerations and for assignment selection. With an inflated system, the Air Force board members couldn't determine who the performers were. The Air Force restructured the OER system from the 9-4 system to a restricted 1, 2, 3 rating system. A rating of 1 (Best) was restricted to 34% of the rated group. 50% of the rated group could be awarded a 1 or 2. A 3 meant you were rated in the bottom 50%. So, if a Wing Commander was evaluating 100 officers, 34 could be assigned a 1, 16 others a 2, and the remaining 50 a 3 rating or less. At the time I called DOD/Joint assignments I had 3 ones. I had received the highest rating for every year. My record was extremely strong. Sargent Browning had pulled up my OER's, reviewed them, and made the 5-assignment choice offer. (Two years after the UN assignment I took the Commander in Chief, Pacific Airborne Command Post (CINCPAC ABNCP) assignment.)

My next assignment established, I still had a year to serve in the Fairchild Command Post.

A Strategic Air Command (SAC) Operational Readiness Inspection (ORI) Inspector General (IG) inspections were no-notice. They just landed no-notice and the inspection began. These were serious events. The unit was evaluated on how well it could generate to nuclear alert status, everyone was tested, and all programs evaluated for compliance.

I was working the command post console and received an early morning radio call from a KC135 saying it was diverting and landing at Fairchild. I called the Wing Commander and told him, "There is a diverting tanker landing and I think it's the SAC/IG." He said, "Thanks" and hangs up. If all goes well, careers continue. Things go bad and there was no forgiveness.

My command post supervisor was away on vacation so I took charge. The Inspector General team landed and the evaluators spread out across the base. I made command post assignments for the controllers.

An inspector appeared at the command post door. The entire wing was recalled as it generated to full nuclear war alert status. The next day the command post personnel assembled and was given written tests. One of the Airman missed a question about command post reports. He technically failed the test. I was sitting with the IG tester and I called the airman over. I asked him a series of questions about reports. He in fact knew about them but didn't correctly understand the IG question. The IG passed him!

An alert KC135 tanker crewmember failed the IG Command and Control Emergency Action Message (EAM) test. My aircrew training program was then closely evaluated. The IG listened to the training tape EAM exam I gave the tanker crews that very week. It turns out, I gave the KC135 crewmembers the exact same Emergency Action Message, and questions, that he failed. So, not my fault.

The Command Post, and my training programs, passed with flying colors. The SAC/IG Command and Control evaluator complimented me on my training programs. He then asked me, "Do you want to be assigned to the SAC/IG?" Another career choice! I thanked him for the offer but said, "No, I have a future UNTSO assignment so don't screw it up."

At my farewell party I was standing in front of everyone saying goodbye. Finally getting to my boss, I commented how great to work with him, and it was, he was a very nice man. Then my last comment, "And he continued to smoke his pipe even though, it was the nastiest, most foul tasting, bitter thing that tabasco sauce could make it." His eyes got big as he realized what I had been doing with his pipe. Then the big smile appeared and he laughed as he realized he'd been screwed with.

The Pentagon

When I left for the United Nations Truce Supervision Organization (UNTSO) assignment I needed to stop at the Pentagon for a few days of classified briefings in the Pentagon, the Air Force, and the Office of Special Investigation. The few days turned out to be several weeks. There apparently were deliberations about a possible reduction of the number of Military Observers in UNTSO. The UN assignment might disappear!

This was my first time to the Pentagon. It was a new world for me. It was massive and getting around was confusing. I was given a temporary location in an AF budgeting organization. I knew nothing about budgeting and didn't want to know anything. I ran down a friend that was in the war fighting tactics development/evaluating organization. It was in the basement near a purple water fountain. I just walked out from the budgeting world and moved in with the war fighters. I'm sure the general I left was totally annoyed. As the days went by the tactic's folks said if the UN assignment cancelled, I could get assigned with them. I didn't like the sound of that.

It was a unique three-week opportunity to observe what transpired in the Pentagon. I didn't like what I saw. Too many generals. Papers were written, briefings given, then papers rewritten, and briefings changed and briefed. There were unbelievable bureaucratic intra agency squabbles about everything. I'm not a bureaucrat, I'm a war fighter! I knew I would totally hate the place.

About week two, I heard that the latest major promotion board results were released. It wasn't my primary year to be considered (my primary year would be the next year), but I wanted to see who I knew that was on the list.

Reading down the list, I knew a lot of captains who were promoted to major. After reviewing the primary zone selectees, I looked at the below-the-zone list. About 5% captains are promoted early, that is, below-the-zone. I see my name! I was selected to be a Major one year early! Wow! It would be about a year before my line number was reached and I actually pinned on the Major rank.

Back at Fairchild AFB, there would be a big promotion party for all the Major selectees. I was staying at

Bolling, AFB. I went to the Officers Club where I found 10 AF lawyers celebrating in the bar. I joined them and had a beer.

More good news. UNTSO was not downsized and I escaped from the Pentagon and departed for Israel. BTW, 30 years on active duty and was never assigned to the Pentagon, I won.

The United Nations Truce Supervision Organization

Here is some background about UNTSO. The United Nation was created in 1947. UNTSO was the first Observer Group it created. It was established in 1948 after the fighting stopped between Israel and the surrounding Arab countries. The purpose of this Observer Group was to observe and report what was really happening from both sides of the armistice lines. These "neutral eyes" gave confidence to all parties involved that the ceasefire was being observed and holding by all parties.

UNTSO Headquarters was in Jerusalem at the Government House. UNTSO was manned by 300 officers from 17 countries. There were 36 from the US and 36 from the Soviet Union. The other countries provided about 15 officers each. Of the US 36 officers, 18 were Army, 6 Air Force, 6 Navy, and 6 Marines.

This was a one-year assignment. UN Observers were usually stationed in Israel for 6 months, then in one of the surrounding Arab countries for 6 months. The Soviet officers were not permitted into Israel because the Soviet Union did not recognize Israel as a country. Their 36 were split between Damascus, Syria, and Cairo, Egypt.

This was during the cold war years, The UNTSO commanders in Egypt and Syria were from the Soviet Union. That meant that US officers, serving in those countries, served under UN Soviet officers. How ironic was that?

There were four new UN members traveling to the Middle East assignment. It was a long flight to Tel Aviv. Before the flight, the four of us met and agreed, for security reasons, not to sit together, not to say anything military, and to put our military ID cards in our shoes. Airliners had been hijacked in the Middle East and military members killed.

We arrived in Tel Aviv late at night and were met by a UN driver with a van. We loaded up and began the drive up to Jerusalem. During the drive the van broke down! Looking at the rear end engine we discovered that the gas pedal linkage broke. We four military minds put our heads together and came up with a fix. We used a paper clip to reconnect the linkage!

It was well past midnight when we arrived in Jerusalem. The streets were deserted as we drove around the old walled city. The walls were lit up, it was an awesome, mysterious, and almost surreal sight to see. We were dropped off at the Palace Hotel in East Jerusalem. It was owned and operated by an Arab family. We stayed there for a few days until we could find a place to rent.

My first morning I took a walk through the streets of the walled City of Old Jerusalem. Wow, it was magic! This was a new and different world. Just outside the wall I watched as two shepherds moved a flock of sheep near the wall so tourists could take pictures for tips. They wore black robes and head gear. I watched as one squatted down with his robe stacking up on the ground. I wondered why? Soon I had the answer. He needed to go to the bathroom! Urine flowed out from under his robe! I wondered what the robe smelled like! A different world, indeed.

Most observers lived in Arab communities on the West Bank. UN members moved around frequently so the rentals were often passed along from one member to another. I rented the lower half of an Arab house in Beit Hinnina, a village on the west bank. It was located between the City of Jerusalem and the Jerusalem Airport.

The Arab husband was a money changer at the Damascus Gate at the walled city of Jerusalem. They had three children and lived on the second floor. I rented furniture. It was a Spartan existence. My first evening there I returned to the house, showered and dressed. There was a knock on the door. I opened it to see the three kids from upstairs standing holding plates of food. The first one said in broken English, "Our mother has sent this for you," they then walked past me through the house into the kitchen and put the food on the table and left.

So, I'm thinking this is a welcome treat. But this went on for four days, food every evening! I'm feeling uncomfortable so I go upstairs. Mom doesn't speak English so I tell the son, "Please tell your mother that her food is just wonderful, and I can't possibly eat all of it, but she doesn't have to cook for me." She responded through the son that, "Your wife isn't here to cook for you so she will." My college degrees were in Sociology so I have studied cultures.

I realized then, that I was in a different culture. The food was awesome. It was wonderful! Lamb and rice, legumes, flatten bread, and a tart yogurt.

My cultural education continued. A few days after I arrived, I put on my backpack and went on a sightseeing walk around Jerusalem. I was in the old walled city walking down the Via Dela Rosa (Way of the Cross.) I met a young Arab man, probably in his late teens. He offered to show me around. We went up onto the roof tops, went down into underground chambers, and walked the Via Dela Rosa. He was very generous with his time so I offered to buy him lunch.

We end up eating in an outside cafe enjoying a great meal. We talked for quite awhile, then as we sat there, he says, "She wants it." I look over at the entry to see a European looking woman standing waiting to be seated. I asked, "She wants what?" He repeats, "She wants it, sex." "How do you know?" I ask. He answers, "Just look at her." She was European and wore shorts, sandals, and had uncovered arms.

A few minutes later I see an Arab girl at the entry. Ok, now it's my turn to screw with him, so I say, "Look, she wants it." He immediately waved his hands back and forth in front of him firmly saying, "Oh no, she is a good Arab girl."

Over my year in the UN, I was assigned to three organizations, Observer Group-Sinai (OGS), UNTSO HQ as Movement Control Officer, and finally to Observer Group-Lebanon (OGL).

My first assignment was to Observer Group-Sinai (OGS). Headquarters was located in Jerusalem at the MAC house. Observers served in the Sinai Observation Bunkers along the east bank (Israeli side) of the Suez Canal. UN Observers living in Egypt were positioned along the west bank of the Suez Canal.

I never deployed into the Sinai desert. When I arrived, Israel had agreed to give the Sinai back to Egypt in three stages. A few nights I slept at the MAC House as Duty Officer and ran the OGS radio network.

Observer Group-Sinai (OGS) was closing down as the Sinai was being given back to the Egypt. OGS observers were being reassigned throughout the area. I was given a new assignment as Movement Control Officer at UN Headquarters. I replaced three Irish personnel. There was a

UN aircraft (F27 Fokker) flown by a Swiss crew that flew out of the Jerusalem Airport. I saw the logic of the assignment: Air Force, so airplane.

I was responsible for scheduling the UN aircraft, preparing the passenger manifests, and was plane side for departures and arrivals working baggage issues. I also worked and coordinated with the Israeli Immigration Officer.

SSgt Gabby, was an elderly (in his 60s) Israeli Immigration Officer. He was born in the walled city of Jerusalem before there was an Israel and had lived through all the wars and fighting. I picked him up and drove him to the Jerusalem airport where he reviewed and stamped passenger passports. SSgt Gabby called me, "Captain Rick." Israelis didn't use last names for security reasons. He was a nice man.

One day we were driving back from the airport toward the UN Headquarters. I needed to stop and pick something up at my rented Arab house. Remember, the house was in an Arab community, and the owners were Arab. I'm thinking, how should I handle this? Do I invite him in? Would the Arabs be offended? If not, will he be offended? I pull up in front of the house. I could see he was really uncomfortable. I said, "Wait here, I'll be right out." I made a quick trip into the house.

After I returned, we sat there a couple of minutes. I could see his mind is racing. I'm interested in what he is thinking so as we sat there, I explained, "I rent the lower floor from the Arab family and they live on the second floor." He said, "This house is like being a millionaire in Israel." I saw where he lived. He lived on the fourth floor of a tenement building. To keep him talking, I said, "When I arrived, they sent me food!" After a few moments he asks, "Did you eat it?" Surprised by his answer I said, "Yes." A few moments later he asks, "Was it good?" I said, "Yes, it was wonderful." Wow! Here was a man who was born in the old Jerusalem walled city and lived his entire life there and wondered if Arab food could be good.

One day the UN plane arrived from Egypt. After all the passengers picked up their baggage and departed there was one small box remaining. The Israeli security officer asked, "Whose is that?" Another US officer, Mark, said he was expecting a package from Egypt. Mark offers to open it up. The security officer responds, "Ok, take it out 100 yards and open it." Mark does, it's the package from Egypt. The Israeli security officer alternative would be to put the package into a hole and blow it up to see what was in it.

Mark had four daughters, I had three sons. In the Arab world having sons made me more prestigious and honorable. Our families were treated differently in many ways. Arab stores were small and focused on only one category of items. There was a meat, a vegetable, and a dry goods store. When I stopped in, they always gave me a glass of hot tea, and I'm sure I paid less when I shopped there then when he did.

As Movement Control Officer, I had a comfortable position at the UN Headquarters and could have stayed there for the remaining year. I volunteered for duty in Lebanon.

My final assignment was to Observer Group-Lebanon (OGL). We moved north from an Arab village to the Israeli town, Nahariyya. It was along the coast just south of the border with Lebanon. John, one of the other Air Force Observers was leaving. He arranged for us to move into the apartment he vacated. It was just next to the beach. Its design was roundish containing about 10 apartments. There was a winding stair well in the middle of the building. Our apartment was on the third level. Every occupant except us was Israeli.

Soon after we moved in, one of Mark's girls' (11 years old) stayed with us for a few days. She and I walked up to the building front door. As I was punching in the unlock door code, she found Israeli money (Shekels) on the ground near the door. It was one note worth about $10 US currency. She was pretty excited. I said, "It must have been dropped by someone that lives in the building, the right thing to do was to try to find the person who lost it." I could tell she wasn't happy with that. Then I added, "If we can't find the owner than you can keep it."

After dinner the two of us set out to find the owner. I didn't know what to expect and thought the folks at the first apartment visited would claim the money. New to the building, I had not yet met anyone else living there. We knocked on the first door and a man answered. I told him that we had found this Shekel note by the front door and asked did they lose it? He called his family to the door, a wife and three kids lined up. He asked them in Hebrew, all said no. They then invited us in for tea! We had a nice talk then moved on to the next apartment. In the end, we stopped at seven apartments and nobody claimed the money, but each invited us in for tea or coffee. I met about half the folks that lived in the building. There were young, old, and even some WWII concentration camp survivors with numbers tattooed on their arms. I let my young guest keep the money.

This experience was truly amazing. My young guest learned a lesson, do the right thing. Yet, she got to keep the money because of the honesty of others. I met most of my neighbors. From then on, we could do no wrong, they all watched over and helped us. I decided that if I ever moved into another country again, I would “find” money, try to locate the owners, and meet the neighbors.

The United Nations Interim Forces Lebanon (UNIFIL) was created a number of years earlier. The PLO were striking Israel from southern Lebanon. Israel attacked into Lebanon pushing all the way up to the Litani River. The UN worked out a deal that if Israel withdrew, the UN would create an armed force, United Nations Interim Forces in Lebanon (UNIFIL), to hold the terrain. This would prevent the PLO from returning to the border. UNIFIL was made up of seven armed battalions from different countries.

The plan called for Israel to withdraw in three phases. The first two occurred as planned and UNIFIL forces occupied vacated terrain. For the final withdrawal Israel inserted the Lebanese Christian Militia commanded by Major Haddad. This strip of land was just north of the border. This area came to be called the “Enclave.” Observer Group-Lebanon had five Observation Posts (OPs) located along the border in the Enclave in Major Haddad controlled area. The three letters after OP were from the nearest village names. They were Observation Posts, OP-Lab, OP-Hin, OP-Ras, OP-Mar, and OP-Khiam.

While we lived in Israel, the OGL Headquarters was collocated located in Naqoura, Lebanon, with United Nations Interim Forces Lebanon (UNIFIL) Headquarters.

I was cautioned that in Lebanon I would meet acquaintances but never friends. There were about 27 armed factions in Lebanon. Their allegiance shifted with the issue. Depending on the issue they sometimes fought each other.

We deployed into Lebanon for seven days at a time. UN Observers lived all around the town and area. Few could read the road signs or addresses so one observer was tasked to be the driver. When a new observer arrived, he hand drew a map showing where he lived. The tasked driver would go to the UN headquarters the day before deployment and copy the maps. He then used the maps to find where the three observers lived so he could to pick them up the next morning.

I served in a variety of duties in OGL. I was a UN/Israeli Defense Force (IDF) Liaison in Metulla, Israel, where two Observers worked UN/IDF issues.

One day the two of us were having lunch in town at an outside cafe. We were across the street from the Metulla police station. While we were eating, a WWII halftrack drove across the border from Lebanon (the Enclave) and parked next to us. I took a few pictures of it. The three-man crew looked to be 16-18 years old. There was a 30-caliber machine gun sticking out the front. I smiled and waved to them, they waved back. They then pulled up across the street next to the police station where they loaded boxes of ammo into the halftrack, I have the pictures. Then they drove back into Lebanon.

A few minutes later an Israeli Captain plus two troopers came over. My camera was sitting on the table. I could see he was concerned about the pictures I took because he periodically looked at my camera. I wondered if he was going to try to grab it. We talked awhile. He immigrated to Israel from South Africa. Eventually they left.

In Tyre, Lebanon, I worked as a Liaison working UN/PLO issues. My PLO liaison was Major Tameraz. We were both born in 1947. He had a map on the wall behind his desk of "Palestine." He said he was born in Gaza but left there as a baby in 1948 and has not been back since. He planned to return some day. I wonder if he is still alive and if he made it back to Gaza. Later on, I think Major Tameraz may have saved my life.

In Tyre, the UN occupied a place called the French Barracks. There were about six old destroyed tanks parked tightly together next to our quarters. The tanks were destroyed in combat because the turrets had twisted holes blasted through them. One day I climbed up onto them. Then I crawled down into one and discover that I could manually turn the tank turret. As I sat with my head sticking out of the tank hatch a few French troops passed by. We talk a minute and then they informed me they were told to stay off the tanks because of possible booby traps. With that, I stopped moving or touching anything and carefully crawled out!

We occasionally patrolled on a road just south of the Latani River. As we drove through an Arab village, we came upon an armored personnel carrier parked alongside a high stone wall. The wall shielded it from being seen from the north, from across the Latani River. We pulled up and parked about 40 feet from it. A crewmember stuck out of the hatch about chest high. He looked over at us, I smile and waved. He said something down into the vehicle. The tracked vehicle moved forward about 20 feet and stopped, he then fired about 10 rounds of .50 caliber machine gun across the Latani River. The bullets ricochet off the stones and ruble on the other side. The target was a place known as, "The Crusaders Castle." This once upon a time castle was mostly now just a massive pile of stones and ruble. There was a PLO position in the ruble. The vehicle then backed up behind the wall. The crewman looks over and gave us a heads-up nod as if to say, "See that!" Here is the irony. There were UN Observers across the river in a bunker with the PLO.

Occasionally a village would report it had received artillery, mortar, or rocket fire. We responded to investigate and prepare a report. The direction of fire can be determined by a crater impact analysis. Mortars fire with high angle of fire and the impact crater very circular. The fuse fin may even be sticking out of the ground at the impact location (Fuse well). The closer the mortar, the rounder the spray pattern. Artillery and rockets are low angle fire weapons and the crater has a more oblong pattern. After we did an analysis and completed a report, we sprayed paint on the crater so the village couldn't claim the strike again.

Once we were sitting in the Tyre Barracks talking with our Lebanese Army liaison counterpart. In came two PLO members who travelled down from Beirut. I wasn't sure why they were there, but in they came. The five of us talked about a half an hour. We served everyone coffee. The two PLO members finally leave. The Lebanese officer then said, "Did you see them watching me to see if I used Israeli sugar?" Our sugar bag sat on the table, it was from Israel and was covered with Hebrew. I wonder what they would have done if he had taken a spoonful of Israeli sugar?

UN training was marginal. We received a few briefings then were turned lose. We relied on fellow observers (the old heads) to show what we needed to do while we were in the field. One day Marine Steve and I drove north from Tyre. He had been there awhile and my senior, a Major, I'm a Captain. We were north of the UN controlled area in the PLO land. A jeep with a post for a machine gun mount passes us going the other way. Steve commented that is illegal and shouldn't be there. He turns around and chases down the jeep. He pulls in front of it stopping it. Steve tells me to stand in front of the jeep. Steve is next to the jeep driver trying to take the jeep keys from him. A small crowd assembles, it quickly grows to maybe 30 or 40.

Soon a Mercedes arrives driven by a PLO Captain. I note that he had a .45 semi-automatic stuffed in the small of his back. He became extremely loud, very animated, and was yelling in Arabic. Neither Steve nor I spoke Arabic.

Standing in front of the jeep, I get a tug on my right shoulder sleeve. It was a small elderly Lebanese man. He says in broken English, "Thank you for what you are trying to do for us but be careful for gun fire from the hill sides." I knew the PLO had occasionally ambushed UN forces before.

This elder Lebanese man was warning me. I yelled at Steve, "Steve, we need to leave, NOW!" We got into the UN vehicle and headed west back toward Tyre. We drove a couple minutes when Steve says, "Guess who's behind us?" I look back to see the PLO Captain Mercedes tucked in close behind us. The road was very curvy twisting around small hills. At a short straight section, the Mercedes accelerated and flashed past us.

A few more curves later and we pass the Mercedes which was parked along the road. The PLO captain was talking with a man next to a house. We pass them. A few more curves and Steve says, "Guess who is behind us again." I look back to see the PLO Mercedes again. A few more curves, then a straight section, and he accelerates past us again and disappears ahead. Further up the road we rounded another curve to find the parked Mercedes loading 3 men with AK-47s.

I said to Steve, "Don't let them catch us again!" He accelerates to maximum speed. We were heading toward the French Barracks in Tyre. But our PLO Liaison (Major Tameraz) was closer. We literally slide into the PLO Liaison parking lot and quickly climb the stairs to his second-floor office. Just a couple minutes later in comes the pissed off PLO Captain who is screaming at us in Arabic. I wasn't sure how this would go down since they were both PLO. Major Tameraz asked us, "Is there a problem?" We say, "No." With a flick of his hand, he indicated to the angry Captain to leave. He left without another word.

I'm pretty sure they would have shot us if they had caught us on the road. In retrospect, we acted improperly. We were not an enforcement organization. We should have just noted and reported what we saw. As I said before, there wasn't much of a training program about responsibilities. Much of what we did was based on judgement and our wits. Steve was senior to me and had been in the area longer so I followed his lead.

I don't believe I was ever personally shot at. However, I did observe .50 caliber machine gun, mortars, and artillery fire. I did miss it at four locations. It seems that incoming fire always occurred after I left a location. OP-Lab took machine gun fire after I left. Tyre took 4 mortar rounds after I left. Naqoura received about 85 rounds of mortar, artillery, and tank fire after I left. Sometime later, Dutch observers died at OP-Kiama. Did I mention the land mines?

The Israelis wouldn't let UN vehicles cross the border. Some years earlier a soldier from one of the UNIFIL Battalions was caught smuggling weapons into Lebanon in a UN vehicle. We drove up to the border, then walked our equipment and food across to waiting UN vehicles.

There were five Observation Posts (OPs) in the Lebanon enclave along the Lebanon/Israeli border. All were pretty much laid out the same. There was a barbed wire fence about 100 feet out around the building. There was a locked entry gate. The building walls were about three feet of reinforced concrete. The ground floor had two rooms, a larger room was the living space/bunk area, and the other room was a kitchen. There were either stairs or a ladder up to the observation tower. One observer was usually in the tower. It had large windows on all four sides, the radios, and a bed. You could go outside and climb an external ladder to get onto the walled roof. Water was hauled in by UN support personnel and uploaded into a large storage tank. There was a bomb shelter under the building. The buildings were painted white with large black UN letters painted on all sides.

We lived in an OP for a week at a time and carried

in all the food supplies for the entire week. Depending on the OP, there were 2 to 4 Observers present.

My first experience with Night Vision equipment was in Southern Lebanon. Events could happen at any time, day or night. Large binoculars worked well during the day. Night vision was another matter. We had a Night Vision hand held sensor. It was about 12 inches long, 8 inches wide and was powered with batteries. There was a button on top. When the button was held down, a whirling sound that could be heard as it powered up. The view through the devise wasn't very clear or sharp. A person's body heat looked like an undefined blob of white matter. Looking through it I could see "Casper the Ghost" like images moving around outside of our barbed wire perimeter. The devise was helpful. (Years later I evaluated night vision goggles for use in the cockpit of the B52.)

Observation Post in Lebanon

The first Observation Post I was assigned to was OP-Lab. OP-Lab was manned with two Observers. It was a communication center and relay station for the OGL radio net. From there we communicated with the other four OPs and UNTSO HQ at Naqoura. Also, whenever a UNTSO vehicle moved around in south Lebanon the right seat Observer held a map and kept his finger on their location. The map was marked up with UN Battalion boundaries, known road blocks, and check points. It had a grid system annotated on it along the top and down the side.

As the vehicle moved around, they kept OP-Lab updated on their location. If they came upon a road block, or approached armed individuals, they would call OP-Lab with the situation and their grid coordinates. Once cleared, they would give us a "clear" call indicating they had passed the road block. This system gave the UN a last known place to start looking for the observers if they disappeared.

I'm on Watch in the Observation Tower

My first post partner at OP-Lab was an Irish officer, Barney Dobby. After the outgoing crew left, he and I sat in the observation tower talking about how we wanted to operate. I told him that under no circumstance would we open the gate for any non-UN person, and we would not leave the compound all week.

He asked why. I said, "Because you're Irish." Then I explained that a few days earlier there had been a firefight between the Irish battalion and fighters from an Arab village. A 16-year-old Arab boy had been killed. The kid had been manning a machine gun. Never the less, the village elders announced that the payback would be the lives of two Irishmen. Barney said that it was only idle talk. I explained we were talking about cultures and it was a real threat.

A few days later we hear a very cryptic emergency radio call for a helicopter pickup for a wounded man. I knew the voice, it was Harry, a US Army officer. Then we heard nothing. We had no idea of what happened or was happening. Later we heard the story. Three UN vehicles loaded with two UNTSO Observers and a bunch of media types drove down from the UN controlled area into the Enclave for a media tour. The three drivers were from the Irish Battalion! Passing through an Arab village they were stopped by armed men. They were all lined up at gun point and the Irish drivers were pulled out of line. Harry and the others were locked in a shed.

The three Irish soldiers realized nothing good was about to happen and broke into a run. One was immediately shot down. When the shooting started Harry and others broke out of the shed. Harry ran to the wounded man and shielded him. The other two Irishmen were being loaded into a car. Harry told the oldest Arab villager that the two UN soldiers needed to stay but the car drove off with them. That was when Harry made the emergency call for a helicopter pickup. Later that day, the two Irishmen drivers were found in a ditch dead.

Most OPs had three Observers on duty. One would be in the tower. The other two would patrol the area in the UN vehicle. We were there for seven days so seven days' worth of food was carried in. We were on our own for breakfast and lunch. Dinner was another issue. Each night a different Observer prepared dinner for the others. It became a source of pride. You wanted to prepare a great dinner for the crew. While I struggled, there were, on the other hand, some unbelievable cooks from Europe! Guys brought table clothes, we shared a bottle of wine, sometimes there was a candle, and always great food.

One evening we were eating and heard artillery fire begin. We scrambled up to the observation tower and started documenting what we observed. The artillery fire lasted for quite awhile. I needed to go to the bathroom so I climbed down the ladder. When I got to the floor, I noted a flickering light on the wall. I looked around the corner to see fire rising from of the dinner table top about two feet high. At first, I didn't understand what it was. Then I realized that a candle that had been stuck in a plastic soft-boiled egg holder had burned down to the plastic. The plastic then burned down into a melted pool of burning liquid! I tell folks I saw the equivalent of the burning bush. The artillery fire eventually stopped.

The Israeli backed forces were always trying to

intimidate the UN to evacuate the OPs because UN observers were reporting on what they were seeing. Occasionally a Militia armored vehicle would smash through the compound gate. The Observers would be taken at gun point, driven to the UN HQ in Naqura, and dropped off. The UN would then return them to the OP.

You never knew what to expect when deploying to a UN position. OP-Hin was under considerable pressure. Headquarters decided that the usual three international Observers would be replaced by four. They added two from the US. I deployed there with Ed (a navy seal), a Belgian, and a Dutchman. When we arrived at the OP the off going three Observers wanted to do nothing but leave. One was totally frantic, he was convinced that Karem was going to come back that morning to kill him! Obviously, they had a rough week. The radios in the observation tower had been ripped out. Some of their belonging taken at gun point. They were warned not to go up into the observation tower. The OP door was gone. Mosquito netting covered the door.

They leave. We had heard that Karem spoke English but didn't with the Observers. Most of the local folks in the area spoke Arabic or French. We discuss how we would handle Karem. We decided to only speak our own country language, plus English, and maybe we will get him to speak English.

I'm talking with Ed, the Navy seal. Ed served in the South East Asia jungles. I asked Ed, "What was your preferred weapon?" He said, "I tried them all but the M16 was best. I've shot people down with a shotgun but saw them get up and run." Ed then asked me, "What is the best way to destroy a B52?" I told him, "Throw an explosive satchel charge on top of the wing near the fuselage. That would blow up the fuel tanks in both the wing and fuselage." I told him about the failed sapper attack at Utapao, Thailand.

Ed walked around all day pumping a dumbbell. Ed said to me, "Rick, they can take my sleeping bag, my food, and my boots, but not this," and he pointed to his dumbbell! He then put it on the floor under the end of his cot.

We get settled in. No radios, no front door. There was only a bug netting over the door. Only three cots, my bed is the kitchen door on two saw horses under the stairs that go up to the tower. So, I'm lying on my door bed reading the book, "The Kissinger White House Years." In comes Karem plus two others all carrying AK-47s. Did you know an AK-47 really looks big when pointed at you? They used the guns barrels to push our stuff around saying, "Motorola, Motorola." They are looking for radios.

One of the guys walked past the end of Ed's cot, stops, and looks down at the dumbbell. Ed immediately picks it up, pumps a couple times, he then extends it out offering it to the man saying, "Do you want this?" Now I'm thinking, if this guy takes it, Ed (the seal) will kill all three! The guy looks at Ed, at the dumbbell, and with a hand waves no and moves on.

As the three leave Ed and I walk with Karem toward the gate. We pass our white UN vehicle. Karem sees the radio in it, opens the driver side door, and tries to jerk it out from under the dash. Ed reaches in and pulls the Mic cord out and gives it to him. Karem hangs the Mic cord over his shoulder. He and I continue to walk toward the gate. He says in perfect English, "So, you're from America?" He speaks English! Wanting to know how well he spoke it I respond with, "Ya sure, ever been there." He didn't answer but continued walking to the gate. I said, "See you tomorrow." He looked back at me, smiled and winked. I knew he was not a physical threat to us. None of us were ruffed up. Karem didn't come back all week! My guess that two US Observers pressured them enough to back off on the Observation Post.

A few nights later the four of us were downstairs talking and we hear the low rumble of engines. We go up into the forbidden observation tower and then out onto the walled balcony. Peering over the wall, we see eight armored vehicles slowly idling past us from Israel north into the Enclave. The road was about 100 yards from the OP. They were moving north to go into ambush/patrol positions.

This is why they (the Militia/Israelis) were putting so much pressure on OP-Hin by harassing observers, taking radios, equipment, and prohibiting observers from being in the tower. Observers there were reporting on this type of activity.

We quickly discuss what, if anything, to do. We had no radios to report this event. Should we launch, pop flares over the vehicles? That would blow their cover. The flares might be seen by them as an attack. They could respond to flares with force, and if they did, we couldn't call for help. We decided to write a report about what we observed for the UN.

Recall that I had my choice of five possible DOD/Joint assignments and I took the United Nations one? One of the other choices was to the US Embassy in Tehran, Iran, flying a C12. A guy who I attended flight school with took it. The year I was in Israel was the year the Iranians overran the US Embassy. There was a scramble for US personnel to get out. He was flying out to listening stations picking up personnel. The story is he flew into a mountain and died. Rumor is that he was shot down. I could have been in Iran instead of Israel.

My tour with the United Nations lasted one year. When I moved to Naharria I bought a BMW 2002 (model, not year) from a departing UN member for $3,000. It was a great car. About two months before I was scheduled to leave, I put "For Sale" signs in the back windows. One day I was in East Jerusalem, driving away from the walled city on Saladin Street. An Arab man runs out in front of me, I stop. He asks me to park and takes me into his small store. He roasts coffee beans. I sat down. He gave me a fresh cup of coffee and requests, "Please stay here for a few minutes while I bring a man here to meet you." As I sat sipping coffee a number of other men entered the store and attempted to approach me. Another man kept pushing them away and out of the store.

The man arrives. He is a businessman from the next street. He wants to buy the car. I said, "I'll sell it for $3,000 US dollars cash, and you pay any Israeli tax. Finally, I said I need to drive it until 15 July." He agreed, we shake hands and I leave.

Two months later, on 15 July I drove up to his store. He was sitting out front with a mechanic. I gave the mechanic the car keys with, "This is the same car you saw before except the back-right window has a slight sticking issue." I'm invited into the man's house to meet his three sons and drink tea. He was curious about the best pilot training schools in the US for his sons.

We talk for awhile, then he speaks Arabic to his son's and I hear the word "Hodish (sp)." I've learned some Arab and this meant something like "how much." I knew he was going to bring up the $3,000 cost of the car and try to dicker me down. He did, saying, "You know the car is two months older so should cost less." I respond, "Yes, but the inflation rate is 120% so the cost should be more!" He slaps his hands together and declares, “We have a deal.” With that he hands me a brown paper bag with US cash in it.

I set the money bag aside and we continue to talk. Eventually he asks, “Aren’t you going to count it?” I respond, “No need to, you’re a businessman, I’m in your home with your sons, the money is right.”

I went with him to the Israeli tax agency. He paid another $3,000 in Israeli taxes for the car. The deal done, I left with my $3,000.

The question then became what do I do with $3,000 in cash? Well, I gave other US member’s cash for their personnel checks. This way they wouldn’t have to pay money changing fees. I got back to the US and deposited all the checks. One $300 check bounced. I called Harry. He had been assigned to the Pentagon. “Harry, I gave your wife $300 for her personal check, it bounced?” He responded, “My check for $300 is in the mail.”

A year earlier I was selected for promotion below-the-zone to Major. I pinned Majors rank on just before leaving Israel. This same promotion board selected me to attend Air Command and Staff College located on Maxwell AFB, Montgomery, AL

The US Embassy coordinated my move. Two guys, who didn’t speak English, arrived at my door. They loaded my stuff directly onto a flatbed truck without any packing. I’m given a manifest written in Hebrew that I can’t read. I watched the truck drive away thinking I’ll never see that stuff again. But I did, and it was professionally packed.

I departed Israel in May 1980 for Maxwell AFB in Montgomery, Alabama, and Air Command and Staff College.

Air Command and Staff College (ACSC)

Air Command and Staff College was considered to be intermediate level Professional Military Education. I was selected to attend by the board that selected me for the one-year early promotion to Major. It was a 10-month school with 550 officers organized into 36 seminars. It was a graduate level program with intensive seminars and lectures. There were 66 foreign officers attending. Officers from Israel and the Philippines were in my seminar.

Academic structure included a core program plus electives. Usually a guest speaker would provide a lecture in the auditorium and then answer questions. We then assembled in our seminar room and discussed the subject. Elective classes were held in the afternoon. There were countless papers to write.

I knew a number of students in the class. One was a pilot from Beale AFB that had been shot down over North Vietnam and became a POW.

In June, 1980, on the first day I was in the in-processing line. This guy and I point at each other and each say, "I know you." But we couldn't figure out how or where. We talked about our various assignments, where, and when we had been places but couldn't figure it out. Periodically we would talk and try to figure it out. Then, about month 9 my light bulb lit. I said, "During survival school in 1970, you and I were in the same element that snowshoed up into the Spokane Mountains for a few days." His eyes lit up and he pointed at me saying, "And you're the guy who was always chopping the firewood." Case closed.

While attending the 10-month school, I worked my follow-on assignment. I called Sargent Browning in Department of Defense/Joint (DOD/JT) assignments. He offered me my choice of three of the assignments he offered two years earlier:

> US Embassy, C12 pilot in Athens, Greece.
> Commander in Chief, Pacific, Airborne Command Post.
> SHAPE Headquarters, Mons Belgium.

When I left Fairchild a couple years earlier for the United Nations assignment, the Wing Commander counseled me saying, "Don't leave the Air Force for too long or it will forget you." His meaning was it could impact the chances of my being promoted again or being assigned to meaningful Air Force positions. I discussed the DOD/JT offers with a number of others. All recommended against taking another DOD/Joint assignment.

It was decision time. I took the DOD/JT Commander in Chief, Pacific, Airborne Command Post assignment. Ironically, some years later DOD/JT assignments became a requirement.

Commander in Chief, Pacific (CINCPAC) Military Assistant

In 1981, I was assigned as the Military Assistant to the Commander in Chief, Pacific (CINCPAC) located at Camp Smith, Hawaii. CINCPAC was a four-star Admiral who commanded all the military forces located throughout the Pacific.

As Military Assistant, I helped plan, coordinate, and support his activities throughout the Pacific. I also assisted with professional and personal arrangements associated with the CINCs official travels. I prepared personal and official correspondence in direct coordination with him. Finally, I travelled with him.

How was I selected for this position? In June 1981, I completed Air Command and Staff College at Maxwell Air Force Base, Montgomery, Alabama. I was assigned as a Battle Staff Operations Officer on the CINCPAC Airborne Command Post (ABNCP) on Hickam Air Force Base in Hawaii. During the cold war, and I'm talking nuclear war, there was network of airborne command and control aircraft around the world on alert that could be launched and, if needed, could direct the nuclear war with the Soviet Union. (They were all were on alert and could take off with just a few minutes warning.)

One EC135 sat ground alert at Hickam AFB. A second ABNCP moved around the Pacific on a 7-day rotation. It was on nuclear alert while airborne. We flew all around moving the platform so it couldn't be easily targeted. It flew from Hickam, to and between, Osan, Korea; Kadena, Okinawa; and Clark AB, Philippines.

The EC135 ABCNP was a nuclear war wagon. There were 28 crewmembers consisting of aircraft crew, communications compartment staff, and the battle staff. It carried an impressive communications capability with UHF, HF, and Low frequency transmitter/receivers. A mile-long cable antenna could be extended to pass messages to submarines. If needed, we could direct the nuclear war.

The current Military Assistant to CINCPAC was due to transfer so a replacement was needed. He was an Air Force officer so they researched all the Air Force officers assigned to the CINCPAC Headquarters. Three officers were considered. I was the only officer to interview with the CINC.

I was initially scheduled to interview with him at

1015 am but our departure from Fairchild AFB, Spokane, WA, was maintenance delayed. We flew the Airborne Command Post back to Hawaii all night from 11 pm to 4 am. We were on nuclear alert inflight, so there was no sleeping. I went home and slept from 6 am until 10 am.

At 3:30 pm I interviewed with CINCPAC! We talked for about a half hour. Then as I was leaving his office he asked if I was a happy man. I stopped, turned around and said, "Yes, I'm an optimist, and life is good!" He selected me.

CINCPAC Airborne Command Post Battle Staff

My first day on the job started at 6 am and ended at 11 pm. I spent the day at the CINC offices at Camp Smith, then I worked a reception in his quarters on the Pearl Harbor Navy Sub Base. I smiled and told them not to take it easy on me. This was a typical day.

About a month later, I was standing behind his desk while he was conducting a promotion ceremony. He was promoting an Army General to a fourth star. I'm standing there and passed out falling across his desk. I was lucky as I just missed landing on the pen stand on his desk that held four upright pens. The CINCPAC doctor looked me over and said I dropped because of low blood sugar. He said I needed to eat better and get more rest.

An Air Force pilot must never pass out! I never used the words, "passed out" with the Air Force flight surgeon. The CINCPAC medical report said I was just dizzy and weak. So, I was dizzy and weak.

I experienced all sorts of memorable events. One event occurred when Philippine President Ferdinand and Imelda Marcos flew into Hawaii. Admiral Long was the senior US official on the island so he met them when they arrived at Hickam, AFB. I rode down to Hickam with Admiral Long.

Security was high. I recall seeing sniper rifle teams on the base roof tops on Hickam, AFB. Ferdinand and Imelda Marcos arrived in separate aircraft, a DC747 and a DC10. I stood at attention at the bottom of his aircraft stairway as he came down the stairs. There was a formal reception ceremony on the ramp with Admiral Long standing next to him.

After the reception, Marcos headed down to his hotel on Waikiki. We drove back to Camp Smith. I asked Admiral Long how it felt to stand next to someone who could be targeted for assassination and did he feel like a target. He said it was not comfortable. Then Admiral Long said we would be meeting President Marcos that evening at the top of the hotel Illikai on Waikiki beach. The top two floors were reserved for the Philippine contingent.

That evening we passed through security and rode the elevator up to the top floor. I knocked on the door. President Marcos opened the door and we entered the suite. There were four other men in the room. President Marcos' and I shook hands and the three of us talked for a minute then Admiral Long and President Marcos moved over to a side table. They talked about US/Philippine issues.

I sat down on a couch with the four men. I don't recall what they all did but one man was a General. He gave me his business card and said, "If you ever have a problem in the Philippines, other than killing someone, call me and I would take care of it." I thanked him.

Now, I have been on the Philippines streets many times and wondered if he knew reality. I didn't know how well his card would be received by any of the locals so later I threw his card away. A few years later President Marcos was thrown out of the Philippines.

CINCPAC and staff took trips west throughout the Pacific and also east to the Pentagon in Washington DC. We flew on an 89th Airlift C135 aircraft. There were mornings on these trips when my alarm clock went off, I didn't know where I was! I'd look at my itinerary and recalled what day it was, which country we were in, and then what the days' schedule was.

There was a trip to Port Moresby, Papa New Guinee, on the schedule. The usual 89th Airlift plane was broken mechanically so an ABNCP EC135 was used for the trip. It was one of the Airborne Command Post aircraft that I earlier crewed.

Papa New Guinee is really a different world. There was an election going on during out visit. There were, in the hillsides, folks sticking each other with spears! Did I mention spears? I walked around Port Moresby a bit. There was a large group (about 30) of locals standing in front of a business watching a TV through the business window. Everyone chewed beetle nut and when they spit it was bright red. Their teeth and spit were a bright red! The ground was gross!

We met with the Papa New Guinee Armed Forces Commander and staff. We had a mega power point presentation describing the massive CINCPAC structure and forces located throughout the Pacific. The Papa New Guinee Commander had two overhead acetate slides depicting his forces. I was embarrassed for him. Their budget was financed mostly by Australia. I didn't know that before.

I've been into many US Embassies around the world before but this was the only time I ever rode in an Embassy car with the US flag flying from the front bumper.

Admiral Long and staff prepared remarks to give to the local media. One comment he would make was that the amount of radiation a nuclear-powered submarine would expose them to would be less than if they were riding in an airliner. I commented that most folks in Papa New Guinee had never been in an airplane and suggested that he say "less than in a canoe." The text was changed to canoe.

On departure day I rode in the front seat with the driver. Admiral Long and wife (Sarah) were in the back seat. There was a departure ceremony planned with a published start time. We were going to arrive 15 minutes early. I had the driver stop and we waited for 15 minutes so we would arrive on time.

Our trips to the Pentagon, Washington DC, were brutal because of time zone changes and travel time. After working all day at Camp Smith, we departed from Hickam in the evening just after sunset. We arrived in Washington around 7 am. The Admiral and wife were in a bedroom compartment on the aircraft where they slept. The rest of us were in seats with work desks between us. We didn't get much, if any, sleep! We arrived in DC with a full work day in front of us.

On one trip I had the opportunity to visit my sister Bev and family. Tim (AF officer) was assigned at the Pentagon. Sister Bev picked me up and took me to their house. Her three daughters were just little youngsters then. I sat down on a rocking chair, talked with everyone for a very short bit, and then I fell totally asleep!

A big deal back then was silk covered beaded necklaces from South Korea, and since we were from Hawaii, pineapples. We brought lots of both to DC and passed them out to all the folks supporting us throughout the pentagon. Word always spread quickly when CINCPAC arrived.

I travelled with Admiral Long to Tokyo, we then flew into Misawa, Air Base, for a visit to the Japanese Air Self Defense Air Force (JASDAF) Officer Training Academy. The JASDAF picked us up and flew us to their Officer Training Academy in an amphibious Albatross! They had two Albatrosses there, one followed along as a backup. We loaded on the parking ramp, then taxied down into the water for a water takeoff. What an experience!

We made a water landing near the Officers Training Academy and taxied up to a dock. Cars drove us through the Academy area where there was not a single person in sight. There were two rows of billeting/barracks with a huge open square area between the two lines of barracks. The area looked like it was made of pea gravel. It was perfectly combed with exact lines across the entire area. An impressive detailed sight! We went to a Museum and attended a presentation that depicted WWII. It was interesting to see the Japanese version of World War II. We attended a few briefings then departed in the cars for the Albatross. The trip back through the open area and billeting building was once again impressive. This time there were about 400 officer cadets in perfect formations at attention as we slowly drove past them.

Our third stop was to Hokkaido, Japan. There was extreme turbulence and heavy rain during the KC135 approach and landing. The arrival was really violent, we were banged around a bunch. We landed! Thank you! The aircraft door opened to a line of cars below the stairs. It was pouring rain. All the passengers ran down the ramp to the cars. We were the last out of the plane. I ran with a Military Aide to a Marine one star general. There was a driving rain and at least an inch of water on the ramp. He slipped, his feet went forward and he fell onto his back. He was totally wet.

We got into the back seat of last car and off we go. As we drove along, suddenly the four cars in front of us turned right but my car turns left! I tried to explain to the driver we needed to go with the other cars. He looked back, **nodded his head, and smiled,** but kept going toward the hotel! I'm carrying the wrapped gift that the Admiral was to present to the Hokkaido Prefecture (Governor) in a planned gift exchange!

Our driver took us to the Sapporo Hotel. So, I'm standing in the hotel lobby with the Admirals gift for Prefecture thinking, I'm totally screwed. The Admirals Executive Officer (a navy Captain) comes into the lobby and grabs the gift from me saying a few well-chosen words and leaves. The gift exchange between Admiral Long and the Prefecture governor was delayed until the Admiral's gift arrived.

The next day we all rode in 3 helicopters (Huey's) for an airborne tour around Hokkaido. I was in number three helicopter with two pilots and a Japanese officer sitting in the back seat next to me. The Admiral and wife were in helicopter number one. Among other things, I'm carrying mementos (CINCPAC pens and key chains) that the Admiral intends to present to his Japanese helicopter crew. I turn to the officer sitting next to me and say that when the Admiral lands he will turn around to look for me because I am carrying mementos for him to pass to the helicopter crew. I need to be there when he lands. I didn't know if the Japanese officer understood me, or if he even knew English, but he nodded his head and smiled. This was my driver's reaction the night before. And I'm thinking there's **the smile and the head nod again** so I'm thinking I'm screwed again!

All three Huey's were spaced out in a sweeping left turn descending to the Landing Zone (LZ). I'm still thinking I'm screwed. Then suddenly the number two Huey breaks to the right and we descend rapidly directly for the LZ. My helicopter lands just moments before the Admirals. As we landed, I was standing on the skid. I think I stepped off the skid before it was on the ground and made it to the Admirals side as he exited his Huey. He looked back for me, I handed him the mementos which he then presented to his Huey crew. Yes! I made it!

Next on the agenda was a car caravan to an outside BBQ luncheon. The entire entourage walked from the helicopters toward a line of waiting cars. I was walking next to the Admiral and his wife. I sensed some unusual activity and movement around us. Suddenly a Japanese man appeared directly in front of me and we all stopped. The man bowed very low multiple times saying something to me in Japanese that I didn't understand. I then realized he was apologizing to me in Japanese. He was my driver from the evening before that took me to the hotel instead of the Admirals meeting with the Prefecture. He caused me to lose face with my boss. They had pre-arranged the entire arrival event so he could apologize to me in front of the Admiral!

I then realized that the Japanese officer next to me in number three Huey totally understood what I said to him. Over the radios, unknown to me, he arranged that the number two would breakout to the right do a 360 turn and land after us. Since he must have known about the problem the night before and the planned apology. He did what he needed to do to get me down in time to get to the Admiral with mementos for his helicopter crew, and not to lose face again.

The line of the cars drove up a winding mountain road to a location for the outside BBQ luncheon. I'm in the front seat with the driver in the lead car. Admiral Long and Sara are in the back seat. Five cars are behind of us. The road twisted and turned between the tree covered hills. After about 10 minutes our driver pulls to the side of the road and stops. All the cars stop behind us. The driver is looking around and has a questioning look on his face like he's not sure of where he is. I look around, there is nothing in the area. I ask him if there is a problem. He smiles, shakes his head no. **There is the smile and a head shake again**! He starts driving again. We came to the BBQ location after a couple more curves. My guess is that he wasn't sure of his location versus the BBQ location.

The BBQ luncheon was great. When we returned to the cars to travel back to the airport and our KC135 for departure we had a different driver.

Being the Military Assistant to CINCPAC opened a few doors for me. I met the 25th Infantry Division Commander. Its headquarters is in Schofield Barracks, Wahiawa, HI. He invited me to ride on an assault landing exercise on a CH-47 Chinook helicopter. The Chinook has an extended body with an engine on each end.

I'm sitting shoulder to shoulder with army troops. I'm wearing my flight suit. They are loaded with a full combat loads. At first, the ride was smooth, but then we suddenly pitched nose down, then the nose jerked up and the ramp in the back end touched down. The forward end remained airborne. Everyone was up and scrambled out onto the ground. Actually, you didn't have a choice. As soon as everyone was out the Chinook rotated forward and climbed. The landing took just a few seconds. Wow, what an experience!

Did I mention I rode the fast attack submarine, the Los Angeles? The Captain of the fast attack submarine, the Los Angeles, was scheduled to brief Admiral Long and staff at CINCPAC HQ at Camp Smith. I helped coordinate the visit. The Los Angeles was docked in Pearl Harbor Sub Base. I phoned down to the boat and a young man answered. Asked him for some information we needed for the briefing. I asked him for the Captains name. He said, "Captain Shitway." What, I'm thinking? I ask him to repeat the name and he repeats, "Captain Shitway." Still not sure so I ask him to spell the name. He spells, "S-H-I-T-W-A-Y." I spell it back to him and he says, "Yes, that's correct." I pass the Captains name onto the CINCPAC briefers for their briefing slides. I remained uneasy with the name throughout the day.

The next day about 45 minutes before the briefing started, I called the Los Angeles again. A different voice answers. I ask for the Captains name. This time his name is Captain S-H-I-P-W-A-Y! I made a panic call to the briefers to change the name from Shitway to Shipway!

The first young man apparently had a B52 crewmember sense of humor.

I met the Los Angeles XO (second in command). He asked if I would like to ride the Los Angeles. I said, "Yes!" He says, "Ok, be on the dock in Lahaina at 6 am on Labor Day. We will cruise back to Pearl Harbor."

The day before Labor Day I rode an interisland flight to Lahaina. I spent the night in an old historic whaling hotel next to the dock. It was a hoot! The floors were rough planks which were uneven. Labor Day morning I had breakfast then went out onto the dock. There were four submarines anchored a few hundred yards out from the dock. A few minutes later a small motor boat docked. I told the young man I needed to get to the Los Angeles and climbed in. He motors me out to one of the subs and bumps up against it. I crawl up onto it and start walking around. Nobody is in sight. I find an open hatch and climb down into the Los Angeles! I'm thinking, "I just entered a nuclear submarine, obviously I'm not a crewmember and nobody has asked me who I was." I'm wearing tennis shoes, blue jeans, a short sleeve shirt and a jacket. I passed crewmembers as I move through the sub. They are all wearing a flight suit type uniform. I eventually walk into the XO who welcomes me aboard.

Fast Attack Submarine the Los Angeles

He led me to the Officers Mess. It was a quite small room. This was where the officers socialized, ate, trained, and briefed. There was a walled in area with a table and enough room on both sides for 3 men to slide in. At the outer end of the table was a single chair, the Captain's chair. Only the Captain sat on it! Meals were passed into the room through a small window.

I met Captain Shipway. He invited me to join him up on the sail to get underway. I followed him up a ladder onto the top of the sail. There were about 4 periscopes on the top. The two forward ones were each raised about two feet. He sat on one, I sat on the other. My guess is that they could go up another 30 feet and I wondered what would happen if the crew below accidently hit the "up scope" switch. How high would we fly? At our feet was a station just big enough for two men. One officer was training the other how to get the boat underway. They called orders to the crew below on an intercom system.

The anchor was stuck. They moved the boat forward, then aft, then forward. The anchor finally broke free and we accelerated faster through the water. As the speed increased water began to cover over the front of the boat. Captain Shipway asked me to go below so he could cycle others up to see the sights. He also told me that when we approached Pearl Harbor for me to join him on the sail for ride through the channel leading into Pearl Harbor. I climbed down into the boat.

What a fascinating world! A few comments. I moved around the boat and talked with many crewmembers. The crew was young, extremely intelligent, and highly motivated. The fast attack submarine was a hunter/killer. They would find, track, follow Soviet submarines, and if required, kill them. Battle stations were exercised. We did angles and dangles! That is, we did a steep angle dive. It looked like we were all standing at a steep 40-degree angle from the deck! I got to look through the periscope. Most of what they do is with sensors and sounds. They knew what each Soviet submarine sounded like! I was surprised when the boat turned. The deck rolled toward the outside of the turn. All my experience was that in a turn you lean into the direction of the turn.

It was a fascinating day. I was standing behind the two crewmembers who were flying the boat. One mentioned that we were 3 miles out of the Pearl Harbor channel. My flying mind put me into a short panic. Three miles is nothing in the flying world! Captain Shipway had invited me to ride on the sail with him through the channel and docking. I will miss it as three miles is a couple minutes air time. Then I looked up at the speed readout, 3 knots per hour. I smiled and realize I have lots of time to get up onto the sail.

I climbed up to the top of the sail and sat with Captain Shipway on a periscope. The two officers are at our feet directing the boat into the channel. There were 20-30 sail/speed boats directly in front of us also lining up to enter the channel. We were quickly gaining on them. Captain Shipway leaned forward and asked the two officers what they intended to do about all the small boats? One officer leaned over and pulls a lever twice that blows a loud horn. At that moment, I saw heads of the boaters all turn back toward us. Immediately all the small boats split turning 90 degrees right/left out of our way! We cruised through between them into the channel.

I completely enjoyed my time at CINCPAC as Military Assistant. What an experience! One day Admiral Long called me into his office and said it was time for me to move on. He thanked me for my service and asked me what assignment I wanted next.

I had been out of the Air Force in the Department of Defense/Joint world for 3 years. It was time to go back to the Air Force. Also, I'd been out of the cockpit for 4 years. I said, "I want to fly the B52s on Guam." He told me to tell the AF General Officer that was on the CINCPAC staff what assignment I wanted next. I picked my own assignment! It's good to have a four-star CINC as a friend.

Two weeks later, I was attending B52D Combat Crew Training School (CCTS) at Carswell AFB, TX.

Chapter 5 – Back to the Air Force

Guam, 60th Bomb Squadron

The training at the 4018th Combat Crew Training School lasted 92 days. I was attending the last B52D initial training class. After we completed the course, the school closed. I had flown the B52D before as a copilot during the South East Asia War (Vietnam). Also, I once was a B52G instructor pilot & check pilot.

The other 17 class members were new to the B52 and new to the Air Force. B52D training began with many days of classes explaining the numerous aircraft systems. The classes were followed by 9 simulators (12 hours). We then flew 15 training flights (119 hours).

Fast forward. Our last flight, 24 March 1983, was a three-ship contingency formation training flight. We led the formation as we returned to Carswell, AFB, for landing. During the approach I thought it would be a historic event if my crew made the last B52D landing of the 4018th CCTS. The Air Force was retiring all the D-models so we were the last class.

We were at minimum fuel but the weather was great. I told the copilot that after approach control changed us over to tower frequency to request a missed approach, to a closed visual pattern, to a full stop. Our following two B52s wouldn't hear our request since they would still be on

Approach Control frequency. I didn't want any competition for the last landing.

We watched the other two B52Ds land as we flew downwind. So far, my plan to land last was working perfectly. Then the unexpected variable occurred. I was about to turn a visual final and land when a 4-ship formation of F4s called initial. Tower told us to extend our downwind and tower would call our turn to base. I knew the F4s would fly over the approach end of the runway and then each would pitchout in turn picking up spacing between aircraft for landing.

B52 Pilot Requalification Simulator Training

I looked down at the total fuel gage, we were now below published minimum landing fuel. Now I'm thinking, wow, what is the chance a F4 would have a problem, stop on, and close the runway! I glanced across the Dallas Metropolitan area and wondered if we could make it over to the Dallas-Fort Worth International Airport if needed. I wondered what impact it would be if we actually landed the B52 there.

Thankfully, the 4 F4s landed and taxied off the runway. Tower cleared us to turn final and land. History made, we were the last 4018th CCTS B52D landing!

Water Survival School

Thirteen years earlier when I was a 2nd Lt, going through the initial B52 training pipeline, I had somehow not been assigned to attend Water Survival School at Homestead AFB, FL. I was now a Major with 78 combat sorties in South East Asia with hundreds of hours flying over the ocean.

Being a Major, I was the senior officer student attending the school. There was a number of "how to" water survival classes covering various survival subjects. We learned how to use the survival equipment attached to us, water parachute landing technique, the under-arm water wings, the one-man raft operation, and the equipment attached to the raft.

Then we spent a couple days practicing in the water. We wore tennis shoes instead of our combat boots so our boots wouldn't be ruined. During this practice they had us pull black stockings up over our white tennis shoes so the barracuda wouldn't strike at our feet thinking our feet were fish.

The plan: The final training day profile consisted of a parasail launch off an Air Force water training boat sailing up to 300 feet. We then would deploy our survival equipment with our survival kit and one-man life raft hanging below us on a lanyard. Once they were deployed, we disconnected ourselves from the towline and would ride the parachute down into the water. In the water, we would pull ourselves up into our one-man raft. Overtime there would be dozens of students bobbing around in rafts. A helicopter would arrive. In turn, we rolled into the water and would practice getting into the helo rescue sling. We would be pulled up out of the water. They then put us back into the water and moved to the next student. Eventually a boat would appear and pick up everyone. Training complete.

Things don't always go as planned. Since I was the senior student, a Major, I went first. I walked out to the middle of the 40 by 40-foot platform on the backend of the launch boat. I was standing in the middle of the launch platform facing to the rear.

My parachute was hanging about 30 feet behind me on a semicircle rack. The launch crew attached me to the parachute risers from behind to my harness over my shoulders onto my chest. The towline from a speedboat was attached to me. I'm told to start running in place. The speed boat will accelerate and after a couple steps the parachute will come off the rack and would lift me into the air. That's the plan.

Oops. The moment I lifted off the deck the tow line went slack! The parachute, off the rack, pulled me up, then back, and slammed me down onto the deck. I landed on the back of my helmet protected head and on my shoulders. It really rang my bell! The three launch crewmembers dove onto me (dog pile) holding me down so the parachute wouldn't drag me off the ships side.

Water Survival School

They disconnected me from the parachute and were all asking me if I was ok? As I rolled over on my side to get up, I saw the line of 2nd Lieutenants who were next. All I saw was a line of shocked big eyes. They were next. I sat up, then stood up, I moved my body around, arms, head, neck, and feeling ok, I gave the launch crew a thumb up and said, "I'm ok, let's do this." They hooked me up again and this time the launch worked and I completed the training.

I was scheduled to fly from Miami back to Hawaii the next morning. When I woke up in the morning, I couldn't lift my head up to get up! Apparently, the head/shoulder landing the day before stressed my neck muscles. I laid there a couple minutes wondering what to do. The room phone was out of reach so I couldn't call for help. Finally, I put my hands around my neck and head. Holding my head, I rolled out of bed and stood up. After a few minutes my neck functioned better and I left for Miami.

Back to Guam

I joined the 60th Bomb Squadron (BS) on Guam on 1 May 1983. I held four positions while on Guam: Crew E-23 aircraft commander & instructor pilot, Squadron Chief of Training Flight, Chief of B52 Standard/ Evaluation Branch

(SO1), and finally, the Director of Training, 3rd Air Division (3AD/DOT).

I arrived in the squadron as a B52D pilot. Because of my previously experience I quickly became an instructor pilot. The 60th BS began a transition from the B52D to the G-model. I had extensive experience in the B52G so immediately completed transition into it. I was both a D & G instructor pilot.

I became the 60th Bomb Squadron, Chief of Training Flight and flew and instructed in both models. However, most of my flying was done in the G-model as we transitioned the D-model crews into it.

The 60th Bomb Squadron was a busy place with many operations and diverse tasking. Being in the Pacific, we flew long distances and interacted with many nations and locations, including Japan, Okinawa, South Korea, the Philippines, Australia, New Zealand, and Diego Garcia. We also sat cold war nuclear alert. Tasked to support the Navy, we trained to drop mines, accomplished Sea Surveillance Reconnaissance flights, and could kill ships with anti-ship Harpoon missiles.

Here are a few sortie profiles:

Takeoff from Guam, then practice air refueling with a Guam or Okinawa KC135 tanker. Fly past Okinawa where USAF & Japanese Self Defense fighters practiced intercepts on us. Continue to South Korea and descend into a low-level route. US and Korean fighters may practice intercepts. Occasionally, we would carry training mines and drop them off along the South Korean coast. We practiced delivering them, their navy would practice mine sweeping. The flight would return to Guam.

Once we flew to Perth, Australia, and flew low level through the harbor dropping training mines. It was a 15-hour flight.

The squadron was tasked to support the Single Integrated Operational Plan (think nuclear war) so needed to remain proficient in low-level Terrain Avoidance flight. After taking off from Guam, an air refueling was accomplished. Flying south, the bomber would descend to a few hundred feet into a low-level route over Queensland, Australia, where Terrain Avoidance was practiced then return to Guam and land. Occasionally, the crew would land in Darwin and spend a day or two, then return to Guam.

One B52D to G transition training sortie became a challenge. I was the instructor pilot in the right seat training the left seat pilot. After air refueling and navigating to Queensland, a descent into low level was begun. The B52G has four engine driven generators. One on each odd numbered engine. One of the four generators shutdown and dropped off line. Training the new guy, I asked him, "What are you going to do about that?" He picked up the malfunction checklist, reads it and says, "We need to do a couple of checks, and if normal, turn it back on with the other three generators." He was correct, that was in accordance with the checklist, I agreed and shook my head, yes. I've seen this problem before. However, this time, immediately when he hit the generator ON button the navigator team yelled over the intercom, "Pilots, what did you do?" There apparently was an electrical surge that fried the entire navigation system!

There were still two more hours of low-level flight ahead of us with no navigation system! We reverted to the navigation basics of airspeed, time, and heading. Our low-level maps had information posted for each leg. The landscape was very isolated, barren turf with few landmarks. Occasionally we would over fly a structure with numbers painted on the roof which I assumed were used to identify them from the air but we didn't have that information. Two hours later, flying time/heading we coasted out only about 2 miles off course. Climbing to altitude, we saw Papa New Guinee off in the distance. We adjusted our heading to pass over the end of it. Then from there changed our heading again to the north to fly us toward Guam.

What a day! We both learned a lesson from this incident. Just because the checklist guided you to take an action, you might not want to take it. Think about the circumstances and situation. The B52G could easily function with only three generators operating. When you're as far away from Guam as we were, I should not have let him attempt to turn the generator back on.

We flew Harpoon Anti-Ship missile training flights. We could carry 12 Harpoon missiles. Two aircraft would fly to a designated area and do a high-altitude radar search. Target spotted, one would go very low, and the other stayed back and high. The high bird would pass targeting data to the low plane. The low plane would be below the horizon so the target ship radars couldn't see it. Harpoons would be launched and the shooter would turn away, still unseen. The missiles flew "stupid" (that is, without its radar on) until a selected time. The missile radars then turned on, would see the target ship, and kill it.

Sea Surveillance sorties were always fun. However, as with any flight, things don't always go smoothly. During the cold war years, we were tasked to locate, identify, and photograph ships for our intelligence community. We would be given a target area to search. Two B52s would fly several miles abreast doing a radar search of the ocean. When ships were detected a B52 would descend and investigate. (We called it bouncing a ship.)

A Recognition Identification Group (RIG) maneuver was flown over and around the ship taking pictures. At 1,000 feet above sea level, we would overfly the ship from the stern to bow. The camera under the aircraft would take pictures down directly over the ship. We then made a sweeping turn and maneuvered to pass across the bow. A pilot would take pictures with a 35mm handheld camera. The next maneuver resulted in a fly down a side of the ship with a pilot taking pictures. We then maneuvered for a pass behind the stern with more pictures. Finally, a fly past the second side of the ship.

Usually we practiced on cargo ships and US Navy ships. Occasionally a Soviet Union combatant was targeted and RIGed. It was the cold war years so it wasn't unusual to have their anti-aircraft gun radar locked onto the aircraft, or, if it was an aircraft carrier, a Soviet fighter alongside!

Flying the RIG Maneuver

One day I was scheduled to lead the two-ship cell out to a search area for Sea Surveillance practice. I was an instructor pilot seating in the copilot seat giving instruction to the pilot in the left seat who was getting G-model transition training. We took off first. Immediately after takeoff our radar navigational system failed making us radar blind. Since it was a clear day, I decided that the number two aircraft would become lead and they could drag us (we would visually follow them) out to the search area which was well east from Guam.

We arrived at the search area and descended to 1,000 feet above the ocean and split-up. We searched visually, they did a radar search. We monitored their location from us by watching them on the air-to-air radio navigation system (TACAN). The system readouts gave us their distance from us. After spending about an hour searching without any ship contacts, we rendezvoused with them using the TACAN. We then followed them again visually toward Guam.

Lead climbed us up to 30,000 feet for the trip back to Guam. We were 2 miles in trail stacked up 500 feet above them. As we flew on, I used the drone time as an instructor to discuss various aircraft systems, and associated what ifs, with the new pilot. Eventually, I asked, "Nav, shouldn't lead be seeing Guam by now?" He said, "Should be anytime soon."

I called lead on the UHF radio, "See Guam yet?" The response was a stunner! He said, "We lost our radar while we were in the search area and are navigating on time and heading." What, I thought! I responded, "You should have told us." If we had known we could have helped them with the navigational problem. We were radar blind thinking we were visually following a fully capable lead.

There was a distinct possibility we could fly past Guam without visually seeing it. There were always low scattered clouds covering the area which could easily hide the island. I look down at the total fuel remaining gage to check our fuel status. There were a couple hours left. We were flying directly west. It was late in the afternoon and the sun was low off the nose on the horizon so celestial shots with the sextant couldn't help us determine our location. Our crew discussed the uncomfortable situation.

A couple minutes later a KC135 that had just departed Guam came on the UHF radio frequency and called out to us in the blind (Just called out our call sign). Our cell lead answered. The tanker said that Guam Center was looking for us because we were late. (Meaning that we weren't close enough yet so the Guam air traffic control long range radar could see us.)

I called lead on the number two radio, "Take a TACAN fix (Distance and radial) off the tanker then ask him where he is from Guam." The navigators could use that data to determine where Guam was located. As a result of the tankers information the cell lead turned us 30 degrees to the right. We flew for about 20 more minutes before Guam radar saw us and our aircraft TACAN locked onto the Guam TACAN ground navigational aid. We found Guam! Had the tanker not called out to us we would probably have flown past Guam to the south.

As you can see the Squadron was heavily tasked with operational commitments. Perhaps the biggest event was the conversion from the D to G model. Three issues occurred simultaneously. Crews transitioned into the B52G, the B52Ds needed to be flown to the boneyard at Davis-Monthan AFB, and finally, B52Gs needed to be flown from the states to Guam.

The B52D and G were significantly different aircraft. Almost nothing, system wise, was the same. About the only thing common from a pilot's perspective was both have 8 throttles. Most young squadron crewmembers had no G experience. They trained initially in the D model at CCTS, at Carswell, AFB. There were 145 D-model crewmembers to train into the G-model.

The Strategic Air Command Headquarters gave the unit a conversion training plan to follow. Each D crew would receive four transition flights in the G, then would go back flying the D. They would then receive one G model refresher flight a month until all the Ds were gone.

I had a couple thousand of G model hours, been an instructor pilot, and had check pilot experience. As the 60th Bomb Squadron Chief of the Training Flight, I told the Wing leadership (including the Wing Commander), that in my opinion, the plan wouldn't work. Four G model difference training flights, then one each month, would not be enough to allow the crews to become proficient enough to pass a 1st Combat Evaluation Group (1CEVG) evaluation. The Headquarters training plan gave us about 5 months to transition the Squadron until being evaluated by 1CEVG. The transition training began.

Concurrently, the D aircraft needed to be flown to the boneyard at Davis-Monthan AFB and the G models picked up and flown to Guam. I flew one of these trips. The plan was to fly a D to the boneyard. Then ride a commercial airliner to Robins AFB, GA. The Robins unit was being deactivated. We would fly 2 G transition training flights back. From Robins we air refueled, flew a low-level Terrain Avoidance route then landed at Mather AFB, CA. After crew rest, we departed for Guam.

The D we flew to the boneyard was ready to retire. We experienced a few issues. During the preflight checklist I accomplished the stabilizer trim check. The leading edge of the horizontal stabilizer moves up or down depending how I move the trim switch on the pilots control column. It's used to adjust the aircraft trim for changed airspeed. Also, when the control column is pushed forward or pulled back the elevators move up or down.

There is large stabilizer trim wheel next to the right knee of the pilot. It spins up or down with any change of trim so the pilot/copilot can see that the trim was changing. Next to the trim wheel was a trim tab indicating pointer. It moved up or down indicating the actual trim setting.

During the check there is a set sequence with the pilot moving it in both up and down directions. There is a crew chief on the ground near the tail reporting over the intercom what the horizontal stab is doing; going up/stop, down/stop, and the final trim setting that was computed for takeoff.

I start the check, the crew chief is making all his calls, which are correct, but my trim wheel in the cockpit, doesn't move! I had never seen this situation before. The trim was working but I couldn't monitor its operation or its setting from the cockpit trim wheel near my knee. I elected to take it as is. I climbed out of the aircraft, walked to the aircraft tail, and visually looked at the horizontal slab trim setting to confirm it was set correctly for takeoff.

We took off and settled in for the long flight. I was in the pilot seat. The sun set and it got dark. The copilot and I reached the overhead cockpit lighting switches turning them on. I then looked down momentarily to my left panel for some reason. When I looked back toward the copilot, I couldn't see him! Smoke was pouring down between us from the overhead panel! I quickly said, "Turn off any switch you just turned on!" The smoke stopped. I then said,

"We will not touch any light switches on for the rest of the flight." There must have been a short. Smoke in the cockpit over the ocean hundreds of miles away from land is not a good thing.

We droned along in the dark. I got out of the pilot seat and put the second copilot in the pilot seat. He was an experienced G copilot who would be getting upgrade to G pilot training flights as we flew back to Guam. I sat in the instructor pilot seat just behind and between the two pilot stations.

After awhile, I noticed the number 5 engine fire warning light element would glow slightly then go dark. This occurred on and off for a long time. The engine fire warning system sometimes gave an erroneous reading. If the warning light actually came on for real, it glowed bright, there was no way to miss it. I watched for awhile wondering if either of the pilot or copilot would notice it, after all, they were flying the aircraft.

Finally, I decided to point it out to them. I didn't want them to see it, get excited, and start to shut down the engine. I said, "Pilot, co, I want to point something out to you but don't touch the throttles." I then showed them the malfunctioning fire warning light. They had not noticed it. It was a system learning experience. As I said earlier, this old aircraft was ready to retire. We landed and turned the plane over to the boneyard folks. I'm sure it has been since chopped up. The remainder of the trip to Robins, then Mather, and finally back to Guam went as planned.

The transition training plan on Guam was completed in the allotted 5 months. In December 1983, the 1st Combat Evaluation Group (1CEVG) team arrived to fly with and evaluate the crews in the G model. To my total surprise, and delight, the 1CEVG team leader was my 1974, "New copilot" from Beale. I hadn't seen him since 1976. He was there leading the organization that would evaluate the unit.

We had a moment alone together and he cautioned me about displaying any knowledge of him. He was worried that it could cause a problem for me within the unit. After dark, he would come over to the house and visit.

The 60th Bomb Squadron Training Flight Instructors needed to fly a check ride as a crew. One event occurred that I had never experienced before. During preflight, the flaps took longer than usual to retract. I had maintenance investigate. There were two motors that operated together to raise/lower the flaps. We used just two flap positions: full down for takeoff/landing, up for inflight. Maintenance found a flap motor circuit breaker (CB) in a wheel well had popped so only one motor was running. The CB was reset. We rechecked it and the flaps operated normal. I said we would take it.

We sat a few minutes waiting for the scheduled engine start time. I mentally ran the "what if" scenarios about what my response would be if the flap motor CB popped during flap retraction. I had never experienced a popped flap motor CB before. Since it, in fact, popped once, I fully expected it to fail again after takeoff. I considered how I would respond. Should I stop the retraction and keep the flaps full down or retract them? If I brought the flaps up and remaining motor died, then a partial flap or flaps up approach would be needed when we returned.

On the other hand, if I kept the flaps down after takeoff, the check flight profile could not be accomplished. We would have to fly around for hours to burn fuel to get down to approach and landing gross weights. Finally, we would need another sortie to accomplish air refueling and low level. I decided that if the one motor quit during retraction, I would continue to flaps up with one flap motor and fly the sortie.

We took off with the flaps in the normal full down position. After takeoff, during flap retraction, we noted the flaps were once again slow to retract and correctly guessed the CB had popped again. I kept the flaps retracting to up position and flew the sortie.

The 1CEVG pilot evaluator watched my every move throughout the flight. He was pretty quiet most of the flight. However, about an hour out from Guam he asked me how I intended to handle the arrival with a flap motor out. I had already thought it through. I said I'd lower the flaps to full down at altitude so we would know the situation before we descended or were on the approach. If the remaining motor failed I could plan for the situation, and fly either a no flap or a partial flap approach and landing.

He asked, "Have you ever made a no flap landing?" "No," I said. He was satisfied with the early lowering of the flaps but said if it failed, he would make the approach and landing since he had made a no flap landing before. I wasn't too enthusiastic about the thought of turning my aircraft over to him but knew, politically, the Bomb Wing would support 1CEVG.

We lowered the flaps to full down at altitude. Issue settled! We descended and flew all the required approaches and landings to complete the check ride. We, Training Flight, passed!

On the other hand, the squadron crews, as I predicted months earlier, didn't do very well. 1CEVG looked for reasons why. They studied the all the Training Folders but didn't find any errors, not even a single discrepancy. Training Flight kept scrupulous records! In the end, they said the crews didn't have enough opportunity to fly the plane enough to become proficient. Gee, I said that 5 months ago about the Headquarters training plan. 1CEVG left, but would return in 6 months, June 1984 to evaluate us again.

The Deputy Commander for Operations called me to his office. He told me I would be moving to the Stand/Eval Division as Chief, B52 Stan/Eval Branch (crew SO1). I argued, to no avail, that I could do more by continuing to train the crews than I could by giving them check rides.

I thought at the time that 1CEVG may have suggested the move to the Wing Commander that I be moved to Stan/Eval. Since then I've asked my former copilot (Team Leader) and he said no, they made no suggestion.

We were now only flying the G model and had 6 months to gain increased proficiency before 1CEVG returned. Training and operational missions continued. I gave crew standardization check rides when needed.

A significant increase in combat capability occurred when the G models were modified with external pylon carriage capability. Prior, only 39 iron bombs could be loaded in the bomb bay. Now 24 additional weapons could be carried externally under the wings. That was big!

We flew exercise, Island Hopper, the 1st live conventional release from external pylons by a B52G. My crew, SO1, led a 7 aircraft formation and was the first crew to drop a full load of internal and external weapons. The Wing Commander rode with us on the IP Seat. The release was from around 30,000 feet onto Farreon de Island. All 7 aircraft released as planned.

The Wing flew in the Korean Theater Large Force

Exercise. (We called it the Big Mac Attack). Over 300 aircraft participated. We were part of B package (Hammer package) which contained 58 aircraft. Ten B52Gs in three cells (3, 3, 4) were flanked by F15s with F4s leading us in. This massive exercise made a simulated a strike on a target in South Korea. It was a show of force for the North Korea.

7 B52G Drop with New External Carry Capability

While I was SO1, a position became available at the 3rd Air Division (3AD). I asked the DO if I could move. He said, "The day after we passed 1CEVG evaluation you can move."

Rule one, always screw with the "Friggen" new guy, the FNG. One time I was on nuclear alert on Guam. There was a new aircraft commander on alert parked next to me. During morning aircraft preflight checks his plane had a maintenance issue. A cockpit "bombs away" light would flicker for no apparent reason. The light should only flash when an electric impulse released a weapon. He called the issue in and maintenance came to his plane to work the problem.

We quickly put a "screw the FNG aircraft commander" plan together. His radar navigator and navigator would jump up and down in the lower deck once causing a jerk in the airframe. The crew chief would call over the intercom, "You dropped one!" and would drop his headset and run. The security police would point excitedly toward the bomb bay. My copilot would point vigorously out his cockpit window toward the bomb bay.

The plan worked perfectly. The FNG aircraft commander thought he had dropped a nuclear weapon on the ramp! That is until he saw everyone laughing.

1CEVG returned in June 1984. The bomb squadron crews had matured by then and did extremely well. As the senior standboard crew (S01), there was a requirement that we receive a complete checkride. It turned out to be a marathon.

On 13 June, we took off on our checkride. Air refueling accomplished, we continued toward Australia for low level Terrain Avoidance evaluation. As we flew along, we lost our radar. Without radar we couldn't fly low level and accomplish terrain avoidance (TA). We turned around to go back to Guam. We were now too heavy to begin approaches so we slowed, put the landing gear down and raised the airbrakes four to increase drag so the fuel flow would be increased. Using more fuel, we returned to Guam just light enough (325,000 lbs) to begin the approaches, then made the necessary landings to complete that portion of the check. We landed. We still needed to fly to Australia and accomplish a TA low level evaluation.

On the 15 June, multiple maintenance problems developed as we preflighted the aircraft. Another crew near us cancelled because of an aircraft maintenance issue. After that crew left their issue was declared fixed by maintenance. Our plane became totally unflyable so I made the highly unusual decision and requested a bag drag to the other "now fixed" plane. Approved by the Wing, we moved and began a new preflight. The second aircraft again developed numerous maintenance issues. I cancelled the flight that day.

On 21 June, we successfully flew to Australia and accomplished the low level Terrain Avoidance evaluation. Our 1CEVG check ride was now complete: Four aircraft, 3 days, and about 20 hour's flight time! Checkride complete!

The next day, I left the Bomb Wing and walked across the street to join the 3rd Air Division as the Chief of Operations and Director of Training, (3AD/DOT).

As 3AD/DOT, I coordinated all B52/KC135 training activities throughout the Pacific. I attended training coordination meetings in the Philippines, South Korea, and Japan. Here is a list of training activities the 60th Bomb

Squadron, CONUS deployed B52s, and Tanker Task Force KC135s flew that I coordinated:

Low level flights across South Korea, and Anti-ship Mine laying operations.
Intercept training for US/Japanese Air Defense Forces fighters.
Live bomb releases on a bombing range in the Philippines.
Low level training across Queensland, Australia, then landing in Darwin, Australia.
Sea Surveillance and photo reconnaissance of Soviet ships.
Tactics Training to launch Harpoon Anti-ship Missiles.
High and low altitude live bomb drops from modified G models on an island north of Guam.

When I was in the Bomb Wing, I led the first high altitude weapons release from the B52G modified with external carriage. As 3AD/DOT, I created exercise "Harvest Coconut." This first-time event was the first low level high drag tactical release using the newly installed external pylons. It also exercised interoperability between Kadena

High Drag Bombs at 300 Feet

F15s and the B52G. F15s deployed to Guam from Okinawa. Crews were given briefings of each other's aircraft and capabilities. Crews then flew for several days with the F15s accompanying the B52s on the low-level attacks dropping high drag weapons on Farreon de Island from 300 feet.

The three B52s dropped high drag weapons in trail formation at 300 feet. The next day one of these aircraft developed a fuel leak during the ground refueling. The leak was from on top of the wing meaning shrapnel came down and struck the top of the wing.

High Drags at 300 Feet

I was invited to Okinawa to ride on a F15 flight as they intercepted the passing Guam B52s on the way to South Korea. After the intercept of the B52s, we fought F15 versus F15. It was a long time since I experienced 9 Gs. I'm pretty sure I was unconscious a time or two.

I was a member of the SAC ADVON team that would deploy to Olson AB, South Korea, during a war with North Korea. The ADVON team would integrate B52 strikes with the tactical forces against North Korea. We exercised the Defense of South Korea twice a year during Team Spirit and Ulchi Focus Lens exercises. During Team Spirit, forces were deployed and practice sorties were flown. Ulchi Focus Lens was only a command post exercise.

I have mentioned occasionally that in my early years in the military I was fighting two wars, South East Asia, Vietnam (Iron Bombs) and the Cold War (Nuclear War Deterrence). But we were also fighting a third war, in Korea. Here's some history:

The Korean War started 25 June 1950 when North Korea invaded South Korea. (BTW, did you know we fought Chinese troops there?) Did you know the Korean War never actually ended? The fighting stopped with a ceasefire when the Korean Armistice Agreement was signed in Panmunjom on 27 July 1953. (I was 6 years old.) The Demilitarized

Zone (DMZ) was created. Since then we have had thousands of troops stationed in South Korea.

I went on active duty in August 1969. In 1976, I moved from Beale, AFB, Ca, to Fairchild in Spokane, WA. I was assigned to the command post at Fairchild AFB as an officer controller. I read all the Operational Plans the Wing was committed to support. One was the Defense of South Korea plan. Our B52s would deploy and bomb assigned targets in North Korea.

Dropping Training Mines along the South Korean Coast

Ever hear about the tree chopping murder in the Korean DMZ? Probably not. On 18 August 1976, a young Army Lieutenant was supervising a crew trimming a tree in the Joint Security Area, the DMZ. The North Koreans attacked. A North Korean picked up an axe that a member of the trimming crew dropped and was used to kill the Lieutenant. The North Koreans claimed that the tree had been planted by Kim ll-Sung.

Ever heard of the USS Pueblo? Probably not. It was a Navy Intelligence Ship. On 23 January 1986 the ship was captured by North Korea. One crewmember was killed, 83 crewmembers were captured. They were abused and tortured for 11 months before they were released.

You better hope that North Korea doesn't get nuclear weapons and ICBM capability.

Years later, in 1983, I was assigned to the B52 in the 60th Bomb Squadron on Guam. We sat on nuclear alert during the cold war years. We also studied targets in North

Korea. We flew weekly training sorties into South Korea practicing bombing and mine laying. We flew annual Team Spirit exercises into South Korea.

On Guam, I became the Third Air Division, Director of Training (3AD/DOT). I have travelled to South Korea many times for various reasons. If war broke out in Korea, I would deploy to Olson AB, South Korea, and integrate B52s into the attacks into North Korea.

The Demilitarized Military Zone (DMZ)

Have you ever been somewhere that was so dangerous the air feels thick? I've been to a number of thick air places in my life. One was when I travelled to the Demilitarized Military Zone (DMZ) Panmunjom, between North and South Korea.

I travelled to the South Korean side of the DMZ. Standing on the south side looking north, I see three rectangular buildings built across a line on the ground that separates the North from the South. Behind me is a US army soldier sitting in a tower who is looking down over the buildings and into North Korea side. He has a hotline phone next to him on the table. The hot line is to call for backup support, if needed. About 300 yards behind his position is a barracks that holds about 30 US army troops. The barracks back door is a platform where trucks are backed up against it. If an alarm is triggered, troops exit the barracks and run into trucks. They are armed with axe handles and/or small arms and they deploy into the DMZ.

Looking across to the Northern side, I see two North Korean soldiers watching us with binoculars from the doorway of a fake building. A number of US army soldiers deploy across the southern area facing the North providing security so we can enter the middle building.

Standing on the North Korean side

There is a negotiations table in the middle of the building with a line across it, North versus the South. Standing on the south side, I asked the US army escort, "Can I go over there on the North Korean side?" The guard did a quick, uncomfortable look around, and then said, "Don't go too far!" With that, I crossed over the line onto the North Korea side, who knew!

Question, when was the last time you stood on the North Korean side of anything? And, I didn't go too far. There were about 10 in the group and I noted I was the only one to do this.

During the Vietnam War I flew 78 combat sorties out of Guam and Utapoa, Thailand. War over, operations shutdown. I don't recall when, but later someone commented to me that there was a B52 capable installation being constructed at a place called Diego Garcia. I never heard of the place again, that is, not until I was assigned to the 3rd Air Division.

Diego Garcia is an atoll located within the Chagos Archipelago in the north-central portion of the Indian Ocean. It is approximately 1,200 miles south of the southern tip of India. The 6,720-acre atoll is horseshoe-shaped with a length of approximately 40 miles.

I've been to Diego Garcia three times. Twice when

assigned as the 3rd AD/DOT, and then years later as the 416th Bomb Wing, Deputy Commander for Operations.

In 1984, on my first trip to Diego Garcia, I accompanied the 3AD/DO (Deputy Commander for Operations) and several other officers on a KC10. We toured and inspected the nearly completed facility. All the buildings, ramps, and runway were state of the art!

The highlight of the trip was a navy helicopter flight around and over the Island. The new runway was lengthened to 12,000 feet. This length made it usable for the B52. Large aircraft parking ramps were constructed. Aircraft refueling capability was built into the ramp. A Weapons Storage Area was created to stock iron bombs. Operations and maintenance buildings had been constructed. A water demineralization plant and a drag chute drying tower were erected. While the facility could support an entire deployed B52 Bomb Wing with hundreds of personnel, there were only 4 NCO caretakers full time at the site. (Six years later, in 1990, during Desert Storm, B52s deployed to Diego Garcia and flew combat sorties from these facilities.)

Diego Garcia, the New Runway

I accomplished a second trip to Diego Garcia. A SAC Operational plan involved deploying three B52s from a base in the US to Guam. The crews rested then departed Guam to fly a low-level training route over Queensland, Australia, they then landed in Darwin. Four KC135 tankers also landed at Darwin.

All the crews rested. Then all 7 planes were launched and the three B52s were air refueled to their maximum gross weight (488,000 lbs.). The two best B52 (maintenance wise) turned west toward Diego Garcia. A KC10 launched from Diego Garcia and refueled the two

B52s back up to maximum gross weight. The two B52s then flew north.

The US aircraft carrier fleet was operating off the Iranian coast. Our B52s would attack the fleet giving them the opportunity to practice fleet defense. The B52s spent about an hour with the fleet then turned east. They flew around the coast of Vietnam, over the Philippines and landed on Guam. The flight from Darwin to Guam was a long 33-hour flight!

As 3AD/Director of Training, I scheduled these flights with the Navy. I was asked by the Navy to travel to the aircraft carrier Carl Vincent and brief the Admiral and staff on B52 capabilities and this exercise.

I travelled to Diego Garcia on the KC10 that would refuel the next deployment. The Navy Viking S2 COD (Carrier Onboard Delivery) is used to transport passengers and cargo to/from the carrier. The COD picked me up at Diego Garcia and flew me to the aircraft carrier Carl Vincent.

It was an uneventful flight. The pilot offered me a copilot seat after takeoff. The flight lasted about 5 hours. He was slouched head down reading a hard cover book and never looked out the window. I guess he sensed my uneasiness with his inattention to outside the plane. He said, without looking up, "Don't worry, it's the big sky theory out here. There is nobody out here."

I left the cockpit for the arrival and landing. I sat just behind the two pilots and could see out the cockpit windows. Boy, did the Carl Vincent look small, real small!

Carl Vincent Looking Pretty Small

We landed and hooked a cable with a rapid stop. I was quickly unloaded and led down into the aircraft carrier.

My first stop below decks was to the F18 squadron ready room. There were about 8 pilots there. I entered and jokingly asked, "Where is the Officers Club?" Well, that didn't go over well with one of the pilots. He ran up to me. I thought he was going to take a swing at me saying, "You friggen Air Force pilots, come out here for a few days." I put my hands up and told him, "You're here because this is what you wanted to do so go friggen off yourself." I don't know what his issue was but let's remember that the fleet had already been at sea 5 months on this deployment.

I was then led to my small cabin by a young Navy enlisted man. As he made up my bed, yes, really, he made up my bed! I asked him how he decided to join the Navy and how was it going. He said, "It was a big mistake, I'm not happy." My room was under one of the launch catapults so was pretty noisy. It was small but at least I was the only one in the room.

Ever been lost in an aircraft carrier? I have, and most of the time! I was toured all over the carrier. I went down into the catapult launch and cable arresting rooms. Both are adjusted depending on the aircraft type and weight launched or landed. I spent time in the tower watching flight operations. Wow, just amazing. It was an intricate dance below me on the flight deck.

Lost on the Carl Vincent

My host invited me into his two-bunk compartment for a lemonade. I know a gin and tonic when I drink one.

I watched many planes launch off the catapult and land catching a cable. As with anything, things don't always work as planned. One time a plane was attached to the catapult. Just before the launch a sonar buoy dropped from the rear of the plane. The launch was stopped as deck crew converged on the plane to fix or remove the buoy. After they moved away the plane was launched. As the catapult flung the plane down the deck I heard three loud bangs. I looked over to my Navy escort and asked, "Did all three tires just blow?" "Yes," was his reply. "So, when the launch hold occurred the pilot probably set the parking brake and then didn't release it for the launch." I ask. "Then when will he come back and land?" My escort said, "In about 45 minutes." About 45 minutes later here it comes, lands, catches the arresting cable and stops.

Usually when a plane landed, after it stopped it would roll back a few feet, dropped the cable, raised the aircraft hook, and then taxied into parking. This one landed but there was no roll back because all the tires were flat. I noted a tug approaching and I expected it to tow the plane into parking. The tug connected and pushed the plane back far enough for the cable to disconnect from the hook. But then the tug disconnected and left. The plane powered up and taxied into parking on all flat tires! A few minutes later the two-man crew walked toward the tower. They didn't look up into the tower. I asked my escort, "So they will be the goats until the next crew screws up." He smiled and said, "Yes."

I toured the hangar deck that was located just below the flight deck. It was packed full of tightly parked overlapping aircraft. It was huge. There was a fire watch monitor in an enclosure overlooking the planes. If a fire started, the watch would punch a button that would cause the entire hangar deck to be filled with foam.

I attended the Admirals lunch and briefed him and his staff on B52 capability. We discussed the current B52 operation. One of the staff officers commented that we burned more fuel on the B52 deployment than they did on a total aircraft carrier deployment. I responded that it took them, the fleet, weeks to get on station while we got B52s there in a few days each with the capability to carry 12 ship killing Harpoons missiles!

While eating lunch, I chatted with the officer sitting next to me. It turns out he was stationed as a Navy Attaché at the US Embassy in Tel Aviv, Israel, when I was there in the United Nations Truce Supervision Organization (UNTSO). I asked, "Did you by chance know the Navy Seal Ed?" Yes, and he tells me a story about Ed. This officer and wife were in bed in their second-floor bedroom in a Tel Aviv house on a Sunday morning. Ed scaled their exterior wall and crawled through their bedroom window to visit! Small world. Years earlier I served with Ed in UNTSO in Lebanon. You read my Ed story already.

There was a non-flying day on the carrier. The non-flying day corresponded with the Soviet Union down day on the Island of Socotra. Remember that this was during the cold war years and they were the nearest threat. The carrier crew did a foreign object damage (FOD) walk. Crewmembers stood shoulder to shoulder and walked the length of the deck looking for trash that could be sucked into

One B52 with 12 Harpoons can kill a lot of ships

a jet engine.

As I walked around the flight deck, I noted there were two fighters on deck each with two-man crews strapped in. My guess was a launch alert. I walked up to one fighter. The back seater was strapped in but hung forward asleep. I told the pilot that I've pulled 7-day nuclear alert in B52s and asked what was he doing? He said, "We are alert 5 and need to be able to launch in 5 minutes. Alert 15 crews are in the ready room, alert 30 is in the squadron." He said alert 5 lasted for two hours, then the alert 15 crew replaced him to become alert 5 and alert 30 became alert 15. If alert 5 launched everyone immediately moved up.

As part of my ship tour I was taken to a lower deck on the aft end of the ship looking out over the ocean. My escort told me, "This is where any crewmember jumped off if they were finished with the Navy." I'm not sure he was joking!

I was in the carrier Command Center during the anticipated B52 attack time frame. Aircraft were picked up on radar coming from the direction of Socotra (The Soviet facility). The watch officer asked, "Are the B52s coming from that direction?" "No," I said. Again, recall that the Cold War was on going and the carrier fleet worried about any plane coming within 300 miles of it. The Soviets had an anti-ship missile with a 300-hundred-mile range! This was what alert 5 was for, fleet defense. Alert 5 launched, and the alert standby crews moved up.

The carrier was developing and testing a new tactic to extend fleet defense radar coverage. They called the tactic the Chainsaw. Two F18s were used. They flew out to the far edge of the fleet radar coverage. One would fly outbound toward a suspected threat direction with its radar looking outward. The other flew the opposite direction, toward the fleet. Then both aircraft would reverse directions. Using this tactic there was always extended radar looking out toward the threat.

My two B52s attacked the fleet as planned. They came in flying at a very low level. The B52s then did fly by

Fly by After Attacking the Fleet

and then departed for Guam. It was a 33-hour flight from Darwin, to the fleet, around South Vietnam, across the Philippines and then to Guam.

The next day I was flown back to Diego Garcia on the COD/S2 Viking. This was my first catapult launch. The launch was a real eye opener! As we were accelerating down the catapult, I was plastered back against the seat back. Once off the carrier I was suddenly thrown forward. That, I didn't expect. In a moment of concern, I glanced out the side window at the water just below us and wondered if we lost engines and are going into the ocean. Not! The catapult acceleration rate is so great that when off the catapult the acceleration rate slows and throws you forward.

I had arrived on Guam a Major (1 year Below-the-Zone). The Lt Colonel Promotion board results was released. I was on it! Why not? I had excellent Officer Effectiveness Reports (OER), I had completed a Master's degree, and all my assignments indicated progression.

I was the 3AD/DOT for 10 months when I received an offer to join 1st Combat Evaluation Group (1CEVG) at Barksdale AFB, LA. I accepted.

Chapter 6 - 1st Combat Evaluation Group (1CEVG)

In May 1985, I moved from Guam to Barksdale AFB, Shreveport, LA.

The Strategic Air Command (SAC) was not a flying club, we were war fighters. During the Cold War years, the SAC main focus was maintaining a nuclear launch and strike capability against the Soviet Union, nuclear deterrence. We also conducted bombing operations in South East Asia (SEA) with iron bombs.

There were two SAC agencies that evaluated units: The SAC Inspector General (SAC/IG), and 1st Combat Evaluation Group (1CEVG). A visit by either organization was a serious event. Do well, because SAC did not forgive failure. Evaluation results reflected on individuals' careers.

The SAC Inspector General (SAC/IG) conducted an Operational Readiness Inspection (ORI) that evaluated each wings capability to perform its nuclear war mission. All wing personnel were recalled. Wing organizations were evaluated for compliance with rules and regulations. Aircraft and crews were generated with nuclear weapons to alert status. Then the aircraft weapons were downloaded.

The unit flew a simulated nuclear war scenario that included a low-level route. Simulated bombs were released and scored demonstrating capability to strike targets.

1st Combat Evaluation Group (1CEVG) had many functions. There was a ground and air component. The ground folks developed low level B52 routes and operated and manned Radar Bombing Sites (RBS) that scored the B52/FB111 bomb high and low training releases. In SEA, they also used radar to direct aircraft in releasing bombs over real world targets.

I was in the 1CEVG air component. We focused on SAC wing aircrew flying ability by conducting aircrew flight checks and evaluating flying related programs. We developed aircrew standard operating procedures (SOP's) and aircraft tactics. We also assisted the ground component in developing low level routes by flying route surveys.

Unit Evaluations

There was a saying out in the field, "1CEVG are like seagulls, they fly in, eat your food, shit on you, then fly way."

We conducted evaluations of SAC B52, KC135, FB111, and KC10 squadrons. A typical Bomb Wing included three Standardization Evaluation (Stan/Eval) crews and 17 Squadron combat crews.

A week after I arrived at 1CEVG, I travelled to Griffiss, AFB, with the evaluation team. I mostly observed the unit evaluation. However, I gave a pilot and copilot a check ride in the Weapons System Trainer (WST) full motion simulator. During the check flight, I gave the pilot a radial and distance designating the point where he was to hold. The pilot screwed it up, he couldn't even fly to the holding point, he failed. I asked the simulator operator to freeze the simulator. I talked with the pilot about where the plane was and where the holding point was. Basically, I explained to him how to navigate to the hold point. I told the sim operator, "Start the simulator." The pilot still failed get to the location.

There are two communication channels in the simulator. Usually, only the main channel 1 was used to communicate between the simulator operator, the pilot team and the instructor/evaluator. When giving a check ride, I always listened to both channels. Here is why. I listened on channel 2, the sim operator was attempting to guide the pilot to the holding point. I told the sim operator, "If you don't shut up, I will fail you!" He didn't say another word. The pilot never made into holding. Then, I asked the copilot to fly the plane and enter the holding pattern. He did. I failed the pilot and passed the copilot.

Later, I led 28-man evaluation teams to evaluate units. We reviewed and evaluated the Wing Stand/Eval programs and Training Flight aircrew training records. Finally, we gave crews inflight and simulator check rides.

We administered two written tests to crewmembers: Bold Print Emergency Action test and Warning & Caution tests. Bold Print was checklist initial responses to emergency situations. Crewmembers were required to list the exact wording, or they failed the test. The Warning and Cautions test require an 80% pass rate.

I had experienced, and survived, a number 1CEVG visits as a unit crewmember, as Chief of Training Flight, and as a Stan/Evaluation pilot. I had developed a personal reoccurring B52 flight manual study program. I knew exactly how many words were in the Bold Print. I would study the warnings and cautions on a reoccurring basis. On Guam, I sat up late into the night studying for these tests to be taken the next day.

As a team chief, I administered these written tests. Occasionally, a young aircrew member turned in their Bold Print test, I would take a quick look at his answers and note he had failed. I would hand the test back to him saying, "Have a seat and review your answers." They would sit down and review their response. Usually they couldn't figure out how to correct their answers and finally turned in the failing answer sheet again.

We gave no-notice flight check rides to squadron crews. The crew would show up at base operations for a weather briefing and find my evaluators waiting to join them. We were obligated to fly with the senior unit Stan/Eval crew. Recall that was me on Guam. I completed a 1CEVG check twice on Guam. The second time: Four aircraft, 3 days, and about 20 hour's flight time! Check ride complete!

Our unit evaluations were pressure packed events. Sometimes emotions triggered conflict. Check ride results impacted crewmember and senior wing careers. Units usually arranged a reception gathering for us. We got to talk with and interact with all the crewmembers. Occasionally, friction would surface from a crewmember.

I had a procedure with my evaluators. If unit crewmembers' emotions got out of hand, I would flag a signal. It's meaning we would go into a two-man policy. This gave us a two person "watch our backs" situation and gave us a witness to any possible event. I only had to institute it once. I ended up pulling my evaluators out of the gathering.

Low Level Route Development

There were low level training routes all around the country. A B52 and/or FB111 could be scheduled to enter a route every 15 minutes. Each year the SAC/IG evaluated SAC wings with an Operation Readiness Inspection (ORI). Part of that evaluation was for them to fly a simulated nuclear war sortie. A big part of the sortie was the low-level route and striking the targets. (See Appendix 6)

The 1CEVG ground component planned and constructed these new routes. At the end of the route they installed a Radar Bombing Site (RBS) to score the weapon release. The site radar locked onto the attacking aircraft. Beginning 20 seconds before weapon release, the aircraft transmitted a steady tone on the UHF radio. At weapon release, the tone stopped. That location was scored by the RBS site. Any score under 90 feet was called a shack (the best). Pause and think about a 90-foot score with a nuclear weapon.

There is a video online (in 2 parts) that explains the low-level route, development, and use. If you look close you might even see someone (me) you might know. Google: "Tone Break: The 1st Combat Evaluation Group Story, part 1", and "part 2."

When all that was in place, a 1CEVG flight crew would fly the new route to conduct a route survey. We checked all the timing/heading data. Most importantly, we looked for any dangerous locations or issues along it. Finally, we flew bomb run a number of times flying lower each time. The purpose was to assure that the RBS site radar could see (radar track for scoring) and communicate with the aircraft. When we were complete, we would fly a "Bug Check" over the RBS radar site. We buzzed them. The RBS personnel expected it and were all outside, on their trailers, and vans. We did a really low and fast fly by. They waved and took lots of pictures.

Flying the Bug Check

1CEVG didn't own any aircraft. Whenever we needed a B52 we would task a unit to provide it. A new route was being developed in Montana. It needed to be surveyed. The Fairchild Bomb Wing was tasked to provide a plane. We flew a commercial airliner to Spokane. We stayed longer than anticipated because of bad weather over the route.

Two days in a row we assembled at base operations to check the weather before going to the aircraft. Weather over the route survey area contained thunderstorms and rain so each day we would cancel. That meant a flyable plane wasn't used for the two days. The third day the wing had a crew planned and ready to fly the plane on a training sortie if we didn't take it.

This route survey was unusual for two reasons. First, we didn't need to fly the entire training route; we only needed to fly the RBS bomb run portion. So instead of entering at the published entry point and flying the entire route, we descended to join the bomb run in visual flight "see and avoid" rules. Second, we would cross an existing low-level route at about a 45-degree angle.

As we crossed the other route, I looked to the left at a sight I had never seen before. Coming directly toward us was a B52 on the established route. It wasn't close. It looked like a V-shaped dash that was trailing black exhaust.

I could clearly see it climb and descend slightly as they flew Terrain Avoidance at 400 feet.

They saw us and obviously were shocked. They had entered on their scheduled time and would not expect to see another B52. I saw the black exhaust increase and the nose rise as the other pilot increased engine thrust. He was climbing to abort the route. He then called out on guard frequency (243.0), "B52 on the low-level route come up on frequency 294.4." We changed over to his announced frequency.

He was clearly shaken seeing another B52 on his route. I explained, "We are a 1CEVG crew on a route development survey flight and not on your route, we just crossed it." We continued and completed the survey. After I landed back at Fairchild, I received a message to call the Minot Bomb Wing Deputy Commander for Operations. He was curious as to exactly what happened.

A few months later on an established low-level route a B1 flew through a flock of geese, ingested birds, lost engines and crashed. SAC had to do something so the low-level route was moved a few miles to the left. Really, like that will solve the bird migration issue? We flew the new route segment survey. John was in the left pilot seat, Tom in the right copilot seat. I'm behind and between them in the instructor pilot seat. John points across the cockpit to the right and exclaims, "Birds." I'm in the IP seat, I see what he was talking about then looked forward. I see wall to wall geese ahead! I point ahead and say, "Watch out birds ahead!" I rolled down to the right and then behind the metal IP seat. John and Tom ducked their heads down below the window and instrument panel. I slowly raised up above my seat back to see clear air in front of us. Thankfully no geese struck the aircraft.

Evaluators/Tactics

When I first joined 1CEVG we were all evaluators. SAC Headquarters was at Offutt AFB in Omaha, NE. Headquarters decided that 1CEVG should have two divisions so half remained Evaluators while the others became Tactics Instructors. I led the new Tactics Division.

I was now in the SAC tactics development world. Our first requirement was to attend the two-week SAC Tactics School at Nellis AFB, in Las Vegas.

The one and only time I had been to Nellis was in 1970 as a student pilot. I flew a T38 there on a cross country training sortie with an instructor. There were topless dancers in the Officers Club! Never saw that before. BTW, years earlier, topless dancing was no longer allowed on AF bases.

Head of the Tactics Division

Twenty 1CEVG instructors spending 2 weeks in Las Vegas, what could go wrong with that? The classes were classified so no material could be taken out of the classroom to our motel rooms for study. The classified classes were extremely informative and, in some cases, a real eye opener. I learned a lot about enemy threat weapon systems, operational capabilities, and their strengths and weaknesses.

The first class we attended, believe it or not, was about the various gambling games in the Las Vegas casinos, how they worked, and what the winning odds were.

I don't gamble, but I thought the class was great because others did. Well actually, I came out ahead $1at the slot machines. To get to the casino Buffets you have to walk past all sorts of gambling games. I saw a one-dollar coin leaning up against the back of a slot machine pay tray and I became $1 richer!

I did gamble a bit years ago in the early 70s while deployed to Guam. During the war, it was a 24-hour operation. There were always card and dice games going on. I didn't play cards but tried dice a few times. The dice game was played with three dice and was called: Ship, Captain,

Crew. One person owned the dice. He laid down any amount of dollars. Going around the table, each person in turn could cover any amount or all of it. Once everyone had an opportunity to bet against the dice holder, he rolled three dice.

If he rolled a 4, 5, 6, he beat everyone and took all the money. If he rolled a 1, 2, 3, he lost everything to the others and passed the dice to the next man. If the dice rolled a pair the number showing on the third dice was the number each player played against.

For example. If the dice holder rolled a pair and a 3, then each man played against 3, so needed to roll a 4, 5, or 6 to win. If he rolled a 1, 2, he lost. If a 3 was rolled, then each kept their bet.

I tried it a few nights and always quit after I lost $20. One night I quickly lost my $20. The dice owner was to my right meaning I would get the first opportunity to cover his bet. He laid down $50. I intended to quit (had lost $20) and said "Have at it" as I was getting up. The dice owner took that to mean I covered the $50 and rolled the dice.

As I was standing there, everyone was looking up at me with a "where are you going" look on their faces. Well, I sat down facing potential $50 loss. Up to this point I had been rolling real bad dice. But my luck changed. I beat his number, won his $50, and took over the dice. I stayed until the game broke up, winning about $350! I don't think I ever played again. That was years ago.

In Las Vegas, on Saturday morning I was leaving my motel room to have breakfast and met three KC135 tanker instructors getting out of a cab. They had been out all-night gambling. One was around $400 ahead. As they walked toward the motel, the big winner asked the other two if they wanted to get breakfast. One said yes and the two of them headed out for breakfast.

The story relayed to me is that they entered a casino but instead of heading directly for the restaurant they sat down at a couple of blackjack tables. After about 20 minutes the big winner came over to the other guy and asked for $50 loan. A few minutes later the big winner came back over and asked the other guy if he would buy the breakfast. You win some and lose some.

The 20 instructors survived 2 weeks in Las Vegas. We learned a lot attending the classified tactics school. Back at 1CEVG, we began to modify the command training program to include information we brought back with us.

There were B52 Weapon System Trainers (WST) at SAC bases. The WST were simulators where the crewmembers could train by specialty or as a crew. There are actually four simulators in a large room. The pilot and copilot simulator replicated the B52 cockpit. It was on pneumatic arms which provided for a full motion. The navigator/radar navigator simulator replicated their stations. Their simulator moved only slightly. Finally, the EWO and gunner simulators did not move at all. Flying as a crew, they communicated over the intercom system. All the instruments operated and interacted as they would in flight.

The quickest way to insert tactics training across the command was with WST simulator profiles. We developed a number of tactics training profiles for the B52 Weapons System Trainers at Griffiss, AFB. Prior to this the WST was only used to practice typical flight maneuvers, low level routes, and bomb runs. We added threats to the flight profiles causing the flight crew to recognize and respond to them.

Another project we worked was a computer program. Have a computer in 1985? Before computers, we used paper manuals to manually calculate aircraft performance including takeoff data, inflight fuel burn, and landing data.

We received a takeoff planning computer program from the field. We ran hundreds, and I mean hundreds, of paper manual takeoff checks against it. We changed the variables: weight, temperature, altitude, and density pressure. We made lots of tweaks to the computer program.

One of our projects was to develop a night vision goggle (NVG) capability for the B52 cockpit. My first experience with Night Vision equipment was in Southern Lebanon in 1980. I was assigned to the United Nations Truce Supervision Organization as an unarmed Military Observer. I was an observer who reported to the UN on what I saw. We lived in Observation Posts and patrolled in Southern Lebanon along the Israeli border.

Night vision technology had improved significantly since my United Nations assignment. We were asked to evaluate the night vision goggles (NVG) for possible use in the B52. The B52 trained to fly low level. The aircraft equipment was capable of 200 hundred feet over the ground!

A cold war tactic to fly low to be under the radar while attacking targets in the Soviet Union.

The B52 had two sensors in the nose to help the crew flying low level. One was a low light sensor and the other was infra-red sensor. The TV like displays had flight instrument information superimposed on them. The NVG would greatly help the pilots see directly out of the plane.

We attached the NVG onto the helmet just above our visors. There was a Velcro patch on the back of the helmet to attach a battery. The goggles could be in the up position or lowered to eye level. With goggles at eye level we could see the cockpit flight instruments by looking down under the goggles.

I'm not sure what the result would have been if we ejected from the plane wearing this equipment.

The first challenge encountered was that the B52 cockpit instrument lighting color did not work with the NVGs. Cockpit instrument back lighting blanked the NVGs out. Either we could see the cockpit instruments or, by turning the cockpit instrument lights off, see through the NVGs. But we needed to see the instruments to fly the plane!

Our initial fix was, believe it or not, a yellow glow stick! We taped a cardboard loop onto the instrument side of the pilots control columns. The glow stick was activated and slid down into the cardboard holder. You couldn't directly see the stick because it was on the instrument side of the control column. All you saw was the light glow on the instrument panel. The cockpit instrument lights were turned off and the glow stick pulled up enough to see the needed flight instruments. We could then see through the NVGs and below them to see the flight instruments. Still, this was not an ideal solution.

The next fix was a Velcro wire harness with little peanut lights. Velcro patches were glued next to each critical flight instrument. We attached the wire harness with Velcro over the instrument panel. There were wires all over the cockpit instrument panel but we could see the instruments better than the glow stick.

The NVGs were awesome! The sight into the night was unbelievable! You could clearly see. It was much better technology than the bulky handheld device I used in Lebanon!

We tested the NVGs for both conventional and

nuclear missions. We flew in low level routes at minimum altitudes and practiced bomb runs. We also attempted to air refuel and to land using the NVGs. I don't recommend using them for air refueling nor landing.

We approved the NVGs for use by SAC units, then we travelled to a number of SAC units to fly with unit crews instructing them in NVG use. (SAC flew B52s low level using NVGs in the beginning nights of Desert Storm.)

There was a plan in the works to have the critical flight instruments (attitude, airspeed, radar altimeter) displayed within the NVGs. That way it wouldn't need to look down below the NVGs to see the instruments.

We developed low level (400 feet) multi-axis bomb runs. The rational for this was to confuse the ground defenses. B52s typically flew in 3 plane trailing formations. The first aircraft might surprise the ground defenses. Ground defense may then turn their guns to the expected direction of the following aircraft.

Three B52s would ingress low level, then approaching the target area, split up. Each attacking the target area from a different direction. Three aircraft dropping a string of high drag iron bombs was tense. Timing was key to this tactic. Miss the timing and they could blow each other up!

The SAC/DO asked us to evaluate if two low level B52s in close formation could cross a target and drop weapons at the same time....at night! He was a fighter pilot who never piloted a B52. We all just looked at each other and shook our heads.

I called the Headquarters project officer and said, "This won't work. Two B52s cannot operate close enough to drop at the same time." He answered, "Yes, I know, but the boss wants us to try, so we need to show him we tried." Our aircraft radar couldn't get us close enough to make a tight formation. The C130 had a station keeping radar (DASK) that they used for night close in formation. I arranged for C130 station keeping equipment be jerry rigged into two Barksdale B52s.

There was no way I would try this at night until we experimented during the day. On the low level (400 feet) I tucked into the lead using the station keeping radar. I had never before flown this close to another B52. Everything worked pretty well during straight flight. Then we tried a heading change turning into our direction. Lead called for a heading change and both aircraft began a roll to the right. It quickly became apparent that I needed to take firm action or be hit by him! I quickly banked steeper and pulled back on the controls. We slid out of the way.

We did a quick chat over the radios about the "That was too close" event. I knew one try would not satisfy the SAC/DO, so I said, "We need to try it again, at least one more time." Lead agreed. We slid into close formation again. As before, at first heading change we once again experienced another near-death event. I called lead, "We are done here, let's go home." A test results report was sent to Headquarters. We heard nothing about the tactic again.

Later, we researched "Emission Control" tactics. A frequency is transmitted when you talk on the UHF radios, operate the forward-looking navigation radar, or the aft facing gun radar. Each system transmits a unique frequency that can be seen out, sometimes, hundreds of miles. Anyone on the ground with a wide spectrum frequency receiver could see our unique frequency transmissions and know we were a B52. If more than one site saw the frequency, they could triangular the B52 location, direction of flight, and direct a fighter intercept on us.

We tested various procedures looking to minimize our emissions. So not to transmit radio frequency at the airport we coordinated with the control tower so they would use a green light to give us clearance to taxi, and then to takeoff.

Inflight, the formation leader needed to communicate with air traffic controllers. The trailing aircraft stayed off the radios but monitored the transmissions and made the assigned frequency changes on our radios so we continued to hear and follow lead.

At high altitude, the gunner flew with his gun radar in standby until we were in a fighter threat area.

During low level the radar navigator normally needed to take occasional radar fixes to update the navigation computer. To do this, a crosshair was run out and was placed over a known ground location seen on the radar scope. He would than "take" the fix. The aircraft computer knew where the ground location "fix" was. The cross hair provided distance and degree information from the known ground location. So, the computer then knew where the B52 was.

Back then, the B52 navigation radar was on the entire flight. To reduce radar emissions, we flew with our radar in standby until just before a turn point. After the radar fix was taken the radar was switched to standby and the turn made. That way, any monitoring ground station that tracked our location and flight direction would look for us in the wrong area.

As tactics instructors, we went to a number of B52 units where we trained tactics including emission control. We pushed the new concepts to the crews and received occasional push back. During a crew flight planning day, we briefed the crew about emission control and why. On one flight, during low level, the radar navigator refused to switch the radar to standby. My instructor radar navigator told me on private intercom. I tapped the pilot on the shoulder and told him, "Tell your radar navigator to go to standby." He did.

We taught all sorts of tactics and maneuvers to unit crews. These events were carefully explained and briefed during mission planning. I also had events that they were not expecting. This was one.

During the return flight to Griffiss, I strapped into the pilot's seat. The plan was for the copilot to fly the enroute descent and accomplish a few touch and go landings. The crew pilot sat watching behind in the instructor pilot seat.

As the copilot pulled power back to start the descent I went on the intercom and said, "Crew we just completed a combat sortie. A surface to air missile detonated near us but missed the plane. It didn't destroy the plane, but shrapnel hit me in the head. I'm going to die soon. Copilot, it's your plane. Use your crew however you want except for the IP seat (his pilot), I will be dead until you land. Then, I will be alive again and will accomplish the touch and go checklist actions. You land, I will select airbrakes-6, I will reset the stab trim to takeoff setting, I will select airbrakes-zero, you will push the throttles up slightly and check engines, if the engines check good, you will push throttles up and make the takeoff. Any questions?" "No," is his answer.

I just dumped a load of pressure on the copilot. This was my first time in a plane with him so I didn't know anything about him. None of the other crewmembers have the pilot's checklists so they couldn't help him with it. He had to run the checklist items, fly the plane, change the configuration, and communicate with air traffic control by himself.

His pilot sitting behind us in the instructor pilot seat leaned forward to help. I gave him a thumb fist pump back indicating to him to back off. The copilot did just great. He talked to the controllers, he put the landing gear down, slowed to flap speed and lowered them, then slowed to landing speed and landed. He landed and I became alive!

Fire Light!

1CEVG didn't own any aircraft. Yet, we needed to maintain flight proficiency. We tasked Barksdale, and Carswell AFB, for a B52 to fly for currency.

This was a perfect weather day to fly. There wasn't any frontal systems or thunderstorms anywhere on the map. The all instructor 1st Combat Evaluation Group crew, accompanied by a Boeing test pilot, was on a proficiency sortie with KC10 air refueling, low level route/bombing activity near La Junta, Colorado, and local pattern work back at Barksdale.

The refueling went well, the flight was smooth and simulated bomb scores close to their best. On the way home, it was decided that I would make the enroute descent and the first touch and go from the copilot (right) seat. The pilot (left seat) worked Air Traffic Control (ATC) on the number one UHF radio. The navigator downstairs passed the maintenance write-ups (things that needed fixing) and the fuel status to the command post on number two UHF radio. Everything looked good as ATC handed us off to Shreveport approach control.

Our smooth, routine flight quickly turned ugly. At approximately 25 miles northwest of Barksdale, we were suddenly vectored further to the north, clear of a C21 that had declared an inflight emergency with smoke in the cockpit. At 35 miles the flaps were lowered and airspeed was reduced to best flare + 20 as approach control gave us vectors for a dog leg to final. The C21 passed across in front of us and I looked to the right watching it disappear in the seven-mile visibility toward the base. The Boeing test pilot sitting in the IP seat called out on the intercom, "Fire light number seven!"

I look back to the instrument panel, the number seven fire light glowed red! I pulled the number seven throttle to idle as I visually looked out to the right wing and engines. Number seven was totally engulfed in flames from top to bottom!

I pulled the throttle to cutoff and then pulled the number seven fire fuel shutoff T-handle out cutting the fuel off to that engine. The fire was spreading. The number eight engine fire light immediately illuminated and number eight was also shutdown. Engine seven and eight are both located next to each other on nacelle number four. I went onto the intercom and said, "Crew we have a serious problem, number seven and eight engines are on fire, check your seats and equipment!"

The possibility to eject was real. Number seven/eight Engine Gas Temperatures (EGT) were rising through 650 degrees to off scale. The nacelle was on fire and there was 12,000 pounds of JP-4 jet fuel in the main fuel tank only feet from the blaze on the end of the strut. The navigator notified the command post of the situation, our intention of an immediate landing, and aircraft evacuation on the runway.

The radar navigator cleared the downstairs crew entry hatch area of all flight gear and equipment in preparation for a fast egress after landing. Meanwhile, the pilot in the left seat declared an inflight emergency to approach control. We were immediately cleared for all headings and altitudes. When you're on fire, people get out of your way.

There were three crewmembers who might have to eject or bailout from downstairs, so I leveled off higher than normal at 4,000 feet above ground level. The two navigators sat in ejection seats that fired down. The Boeing test pilot would manually jump out of one of their ejection seat holes.

Although the fuel shutoff valves were closed, the fire did not let up. Flames were blowing straight back in the airstream, but there was no indication fire was moving up inside the strut toward the wing fuel tank. I watched the strut for indications of becoming dark, an indication the fire was spreading up through it. A fire in the wing would mean evacuation the aircraft be required.

At this point I had three possible actions to consider. I could accelerate the plane by retracting the flaps and landing gear and the engine fires might possibly blow out. Or, there have been instances where burning engines fell off the plane, or I could race to the runway and land. I decided to land because we were configured to land (Airbrakes position four, landing gear and flaps down) and there were fire fighters already waiting for us. We were only seven minutes from the runway.

I increased the airspeed to 185 knots (210 mph) in an attempt to possibly blow the fire out but more importantly to reduce flight time to the runway. The radar navigator had the radar crosshairs on the end of the runway that provided the time and heading to touchdown. The direct course to the runway was well right of the normal approach but avoided populated areas. I made a couple of heading changes to avoid flying over a few apartment complexes.

As I was flying, the other pilot glanced around the cockpit and we reviewed the affected systems. Number seven hydraulics read zero, which meant the right outboard spoiler/airbrake group would be inoperative. However, control problems associated with the spoiler/airbrake group were negated by the effects of two inoperative engines on the same side. Number seven generator was off, but its load automatically was picked up by the other three.

A six-engine approach and landing, or an asymmetrical thrust go-around at 275,000-pound gross weight with good visibility, was controllable and should not be a problem. The problem was the fire and potential problems caused by the preceding C21 emergency landing just two minutes ahead of us. If it was stopped on the runway, we could not land. That would be a problem.

I reviewed the crew evacuation plan: stop on the runway, shut down, egress as quick as possible, the nav team would throw two extra parachutes in front/behind the front landing gear, then we would all assemble at least 1,000 feet up the runway from the nose of the aircraft.

I elected to fly a visual approach, using the Instrument Landing System (ILS) glideslope indicator as a backup. Up to now the offensive team stayed off the intercom, coordinating with command post on number two radio, letting the pilots concentrate on flying the aircraft, but now the navigator called for a configuration check requested by the command post. They then relayed the pilot's confirmation (gear down, flaps down, and airbrakes 4) over the radio number two.

As I started the final descent, I slowed to 135 knots (155 mph), and re centered the rudder trim. One last configuration check—gear & flaps down, airbrakes 4, lights on. The 20 seconds to go seemed like 20 minutes. There were lots of flashing lights on the ground along the runway. Emergency vehicles were everywhere. Then touchdown.

The airbrakes were raised to airbrake six, I deployed the drag chute, and stood on the brakes. The navigators started turning off all navigational and oxygen equipment downstairs, permitting the navigation system spin down on battery power.

Tower confirmed the engines on number four nacelle were still on fire: "ALGER-64 (our call sign), fire appears to still be burning, ALGER-64 recommend you evacuate, evacuate! The fire is spreading!"

Fire trucks were charging in from every direction. As the aircraft came to a stop, the navigators called for an open window before opening the entrance hatch. "Everybody out!" I announced as I activated the evacuate red light, but no one needed any encouragement.

A parked KC10 had burned and exploded on the ramp just a few days before! The memory of a burning aircraft was still fresh in our minds.

I set the brakes while shutting down the remaining six engines. I then shut off all the fuel valves on the fuel panel and finally turned the aircraft main battery switch off.

The Boeing test pilot laid my number one ejection seat safety pins on my leg and left. The pin was my way to safe my ejection seat. I attempted to unstrap from my parachute and ejection seat. As I flipped both parachute leg strap buckles, the right one would not release and I struggled with it! "You coming?" the other pilot yelled from the exit ladder, then he disappeared down the ladder.

I semi stood up and finally shook the parachute straps loose. Reaching the ladder, I looked down through the lower deck to the ramp. I saw a fireman's gold face plate and silver covered firefighting suit squatting on the ramp looking up into the plane. I had two quick thoughts. First, I felt a sense of relief knowing that the firemen were there taking care of us. Then, I thought, compared to him, I wasn't dressed to be in an aircraft that was on fire!

I quickly climbed down the ladder and exited the hatch onto the ramp. I was the last one out. I squatted under the plane on the ramp face to face with the fireman. I held up my five-finger hand spread. I yelled, "I'm number five…I'm the last one, nobody else is in the plane!" The silver helmet nods and I get a "thumbs up." Telling him the aircraft was empty meant he didn't have to enter a burning aircraft looking for crewmembers.

The navigator team led the way as the crew ran up the runway away from the burning B52. By now the aircraft engines were being smothered with white foam and looked as if it had just been caught in a huge snowstorm. When the fire was out, we could see the "spaghetti" of wires and tubing on the sides of the engines because the metal cowlings had totally burned and melted away.

Part of the burning metal fell from the aircraft and ignited a number of grass fires north of the field. The local fire departments responded and quickly extinguished the fires before any significant damage had been done. I was glad I didn't over fly the apartment complexes!

Things can change rapidly in the aviation world. The entire episode lasted just seven minutes from fire light to last man out of the aircraft. This is the aviation world. You never knew what would happen in the next second.

We Landed with those Two Engines on Fire

It was determined that the number four nacelle (engines 7 & 8) fuel line cracked, due to vibration. The crack allowed JP-4 jet fuel to spray over and around the two engines. When we slowed down to configure for landing the fuel torched off.

While at 1CEVG, the Colonel (06) promotion board results were released. I was promoted one year early (one year below-the-zone)! One thousand ninety-nine Lieutenant Colonels (05) were selected to be promoted to Colonel (06).

I was one of 156 selected 1 year below-the-zone, 3.6%. I recall once hearing that for every 100 new Second Lieutenants entering the Air Force, only 4 would ultimately be promoted to Colonel. I was also selected to attend a Senior Service School at the National War College, Fort McNair, Washington, DC.

After the promotion results were released, I received an assignment as Assistant Deputy Commander for Operations at the 416th Bomb Wing, Griffiss AFB, Rome, NY. I would travel there in March 1988.

Chapter 7 - Griffiss Deputy Commander Operations

I was initially assigned as the Assistant Deputy Commander of Operations (ADO). After about 6 months the DO transferred and I became the DO.

Eighteen years earlier I was a 2nd Lieutenant copilot who just completed initial B52 flight training. I was standing on a slightly raised platform briefing the Wing Deputy Commander for Operations (DO) on a nuclear war SIOP (Single Integrated Operational Plan) sortie. When I finished, the Colonel grilled me with a lot of questions about the mission. When he finished, the 20 staff members sitting behind him asked me still more. The DO finally raised his hand and said, "That's enough, he knows the mission, I will certify him." With that I became a SAC warrior! Now I would be the DO asking the questions!

How did I get here, this assignment, the DO? A little over a year ago I was in 1CEVG. Throughout history 1CEVG was the SAC organization that evaluated units flying capability. SAC decided to split the organization into two divisions, Evaluators and Tactics. I was selected to be head of the Tactics Division.

I had traveled to Griffiss AFB a number of times. One week after arriving at 1CEVG we travelled to Griffiss AFB to conduct a unit evaluation. I went along to watch how 1CEVG evaluations operated.

Then 1CEVG changed and I moved to lead the Tactics Division. The first unit we travelled to was Griffiss, AFB. My tactics instructors flew with the crews training tactics and the use of Night Vision Goggles.

I debriefed the Wing Commander and his assembled staff about our visit. I said, "First, I want to thank everyone for the hospitality shown to us. The change within 1CEVG was historic, we are in the Tactics Division and as such this was not an evaluation, it was training. At the end of the year we will make a genetic report about the command tactics strengths and weakness we note throughout SAC units."

After my debriefing, the Wing Commander invited me to his office. The Vice Commander attended. The Wing King asked me for my candid opinion of his crews. Since he asked me a direct question, I gave them my candid observations, the good, bad, and the ugly.

Later I traveled to Griffiss a number of times to develop Tactics Weapon System Trainer profiles that would be used across the command.

I came out on the Colonels below-the-zone (a year early) list. The Griffiss Wing Commander must have liked what he saw, because he asked to have me assigned there.

This was one awesome assignment. Now mind you, there isn't a school to teach you how to be a DO. I'm not sure how other DOs operated. You're on your own. It brought together all my prior experience base. I put to use my B52 operations experience, command and control experience, evaluation and tactics knowledge. And best of all, I was still flying as a B52 instructor pilot. It was unusual for a wing to have a DO who was an instructor pilot.

I remember years earlier when I was Captain Instructor Pilot at Beale. The Colonel DO would call me frequently asking me my opinion about what a crew did during a training sortie the day before.

What was also unusual, I was still a Lieutenant Colonel. I don't recall I ever saw a Bomb Wing/DO that wasn't a Colonel. I spent my entire tour at Griffiss as a Lt Colonel. I didn't pin on Colonel rank until after I had arrived at the National War College.

The Wing Commander was the senior officer in the Bomb Wing. Four senior officers supported him in these functional areas: Operations, Aircraft Maintenance, Base Commander, and Resource Management.

As DO, I oversaw 380 persons organized in eight divisions and two tactical flying squadrons (B52G/KC135). We operated with 19 B52G and 16 KC135 aircraft. I supervised our nuclear alert aircraft and facility, the command post, aircrew flight/ground training schedules, current operations planning, the nuclear war planning vault, the simulator, and the base operations.

Long ago as a command post controller I called the

DO day and night with operational issues. Now I was receiving these calls. I had hot line phones next to my bed, in the kitchen, and in the office. I had a UHF radio in my car. And I also carried a hand-held radio that was called a brick (it looked like a construction brick.)

My Assistant (ADO) did a lot of routine office paperwork for me. I didn't like to be in the office, rather, I wanted to interact with the troops. I did a lot of "walk abouts." I'm sure I made my division chiefs and two squadron commanders uncomfortable, but I would showup unannounced in the squadrons and divisions. I'd walk around and talk with the troops.

When I first arrived on base, I went to the bomber alert family building. This building is located just outside the B52 alert area where families met and spent time together. They didn't know who I was. I talked with many crewmembers and wives. I learned a lot about the unit culture.

I was a bomber guy in a Bomb Wing. The Wing included a tanker squadron. Typically, I think the tanker community felt like second class citizens. I knew different. I had learned to appreciate the tanker mission and how important the tanker world was years ago. Nothing moves without tankers.

I needed to get their attention and show the tanker community that I viewed them as important to the wing mission. After looking over the flight/ground schedules, I picked the event.

At 1 am, I drove out onto the tanker parking ramp, parked the car, and crawled up the KC135 crew entry ladder. The tanker crew was obviously surprised, I mean shocked. I was so new to the wing that I wasn't even sure they knew who I was. I said, "Please call the command post and tell them to add Charlie (my call sign) onto the flight orders. Tell them that the ADO will be taking my calls." I already knew the answer, but asked, "What are we doing tonight?" The aircraft commander responded, "We will refuel a C141 that is flying to Europe." I knew the wing was tasked to support these reoccurring C141 flights crossing the Atlantic. It was a great flight, I totally enjoyed it. I interacted with the crew. I laid next to the boomer in the air refueling pod and watched him refuel the C141. I've been at the other end of the boom many times! Later, I was told that I was the first DO to fly on this flight. Word of my no-notice tanker ride spread rapidly throughout the tanker squadron! Hopefully, I got tanker community attention.

My flight experience was in the B52. I had just come from 1CEVG and was still a B52 instructor pilot. I knew the plane inside out. I needed to learn the KC135 because I would be getting all sorts of calls about airborne and ground system problems. While I had ridden on the KC135 many times, I needed to know how the aircraft systems and the flight operations worked.

I was sent to Castle AFB, Merced, CA, for KC135 training. I attended numerous system classes and flew numerous simulator training flights. I returned to Griffiss and completed a flight check. I was now a KC135 pilot. This training proved very valuable.

Alert Christmas Tree

I was fortunate to have extremely capable commanders in both the Bomb and Tanker Squadrons. Also, all eight Division Heads were absolute experts in their fields. They all gave me total support.

My first mission priority was to support the Single Integrated Operational Plan (SIOP), as in nuclear war. There were nuclear loaded bombers and supporting tankers on alert needed to be able to takeoff, within 15 minutes notice!

There were a lot of moving parts in the Bomb Wing. Crews needed to be scheduled and trained. There were ground and flight training schedules to produce. Evaluators tested and gave checked rides. Nuclear War flight plans needed to be constructed. Alert facilities were needed to house and feed the crews. Command Post was the focal point for all wing activities.

When the klaxons blew, alert B52 and KC135 crews raced to their aircraft, started engines, and prepared to taxi. They were tasked with being able to takeoff within 15 minutes. During the response crews copied and decoded an Emergency Action Message (EAM) that would direct their actions.

Griffiss B52 Squadron

I arrived at Griffiss AFB, March 1988, as the Assistant Deputy Commander of Operations. I didn't leave the base for three months except as Deployment Commander to Elmendorf, Alaska; Salina, Kansas; and to Diego Garcia.

Elmendorf, Alaska

From 18-30 April 88, 3 B52s deployed to Elmendorf, Alaska, on exercise AMALGAM WARRIOR 88. I was scheduled to ride a C141 with the wing maintenance and support troops.

As I approached the C141, I noted a spirited discussion going on near the crew entry hatch at the front of the plane. It was between the C141 loadmaster and a couple of my people. I walked up and asked, "What's up?" The loadmaster said, "There isn't enough urinals on board for the number of personnel we were loading. "No problem," I said, and called the command post on my brick, "Please have a wing urinal delivered to the C141." I signed a hand receipt for a urinal. I guess that makes me a problem solver. The urinal arrived and we departed in the C141 for Alaska. I don't know if the urinal made its way back to Griffiss. I'm still waiting for a bill for it.

There were about 300 aircraft in this exercise. The exercise occurred in Stoney Military Operating Area (MOA). Alaska is big and totally beautiful!

The exercise plan called for two attacks into Stoney MOA each day, one in the morning and a second in the afternoon. We, 3 B52s, were aggressors that flew low level to attack planned targets with conventional weapons. There were fighter defenders with AWACS assistance who defended against us.

We flew different aircraft and lived different work hours. We took off and flew the first Gorilla (a TAC term meaning mass attack). Then the tactical forces (think fighters) landed for lunch. While they lunched, we flew a low-level training route, then returned for the afternoon second Gorilla.

One morning, the weather was too bad for the TAC fighters to fly. They cancelled. SAC B52s took off and flew the morning Gorilla route alone, then flew the low-level route, then joined the TAC fighters for the afternoon Gorilla. At the debriefing I decided to screw with the TAC community and made fun of them for not being able to fly.

"SAC aircraft took off, and since the weather was too bad for TAC to fly we changed from a conventional war to a nuclear war scenario. We up loaded a nuclear bomb and missiles load." Using overhead acetate slides, I depicted launching an 18 nuclear missile gorilla, then descending to the low-level route to release 12 gravity weapons. The room was totally quiet. I could see amazement/shock on the faces of many of the fighter pilot faces. I got the feeling many of the fighter folks had no idea of the firepower we could carry.

My previous assignment was a Tactics Instructor in 1CEVG. Tactics in war means "anything goes" to win. On one flight, we had a gunner use a common whistle to jam the F15 enemy fighter communications. We figured out what UHF radio frequency the F15s were using to communicate and coordinate on. It was dialed into our number two UHF radio. Every time the F15s or AWACS began to transmit the gunner blew the whistle over their frequency. The F15 couldn't communicate. Just think what it would be like to have a headset on and someone blowing a whistle into your head! Post flight there was a mass debriefing and critique. Boy, the F15 fighter community was pissed! And I mean, in my face pissed. I responded, "If it was that easy to neutralize you guys, you had better hope the real bad guys don't hear about this."

There were Canadians attending and flying the event. One approached me and we brainstormed how we might work together. We came up with a tactic, a plan. He flew a T33 (Think really old fighter). He would join up with us at a planned location as we flew low level. He tucked into close formation with us where the AWACS radar would only paint one aircraft. When a defending fighter attacked us, the Canadian in the T33 pulled up totally surprising the F15 pilot. It worked! A Canadian in an old T33 shot down a state of the art F15. Tactics!

The exercise over, I planned our departure. I wanted to show the TAC forces, and Elmendorf personnel, something they probably had never seen before. I briefed the crews that we would fly a Minimal Interval Takeoff.

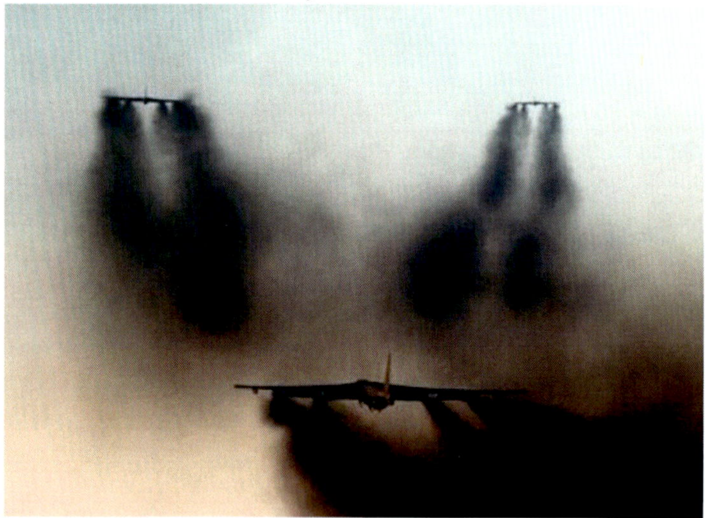

Minimal Interval Takeoff

Three 52s taking off with a 12 second interval between them. Three B52s on the runway taking off at the same time.

Salina, Kansas

From 28 July-11 Aug 88 I deployed as Deployment Commander to Salina, Kansas, on MIGHTY WARRIOR 88. Historically, SAC operated from well-established bases. This exercise was planned to learn how to deploy and operate from an austere forward operating location, a bare base. The biggest challenge would be the maintenance, support, and logistics organizations. They needed to plan what to bring, and how to replenish expendable items. Jet fuel was delivered daily by line of tanker trucks that down loaded jet fuel into huge rubber fuel bladders. Drag chute folks needed to dry and to repack the drag chutes. They didn't have the heavy-duty equipment that existed back at the base.

Massive Fuel Bladders

Over 400 personnel from 7 bases converged on Salina. I had never been there before. I felt more like a Mayor than Deployment Commander since I was involved in everything, and I mean everything, not just operations.

To get organized, I established a deployment staff meeting every afternoon. All senior members from all the various functions were represented. About 35 attended. I gave everyone an opportunity to address the assembly and explain their functions and comment on problems and issues.

We lived in tents, an old movie theater, and in an old National Guard barracks. Flying night sorties meant most people slept during the day in 95-degree heat. Chiggers were a big problem!

Ever feed 400 plus people? A mobile food trailer and cooking crew was transported in. We served 4 meals a day: Breakfast (Hot Meal), Lunch Meals Ready to Eat (MRE), Dinner (Hot Meal), Midnight Chow (MRE).

After three days of eating, I was totally underwhelmed by the food that was being served. Cooked food was cold, under cooked, or whatever. 400 plus people working a 24-hour operation in extreme living conditions deserved a great meal. I called in my Resource Manager and told him, "I need an experienced cooking crew here. These folks aren't hacking it. The troops deserve better." There was a new cooking crew in the next day. Problem solved.

At one early staff meeting the head cook said that people were entering the food supply tent and taking food. We all discussed the issue for awhile. Apparently, our schedule caused some troops to miss a meal time. Finally, I said, “Please pass the word to all the troops that I will court martial anyone found in the mess supply tent. On the other hand, anyone showing up for one of the 4 meals will be feed.” The head cook responded, “But everyone is only authorized 3 meals a day.” I repeated, “I’m aware of that, but as I stated a moment ago, anyone arriving to eat will be fed.”

I did a walk around with the senior NCO’s. The old movie theater bathrooms were totally trashed. I suggested that a duty schedule be setup to clean the bathrooms.

We flew 6 B52 sorties each night. Late afternoon on the first fly day, one of my staff members came into my work area and told me, “There are about 300-400 people outside the aircraft parking ramp entry gate. I lowered and slowly shook my head. After experiencing Vietnam antiwar protestors against the military years earlier, to me this meant they were probably protestors. I said, “Ok, let’s drive out and see what’s up.”

Here was my problem. There were security police with M-16’s (no bullets) at the ramp entrance gate, but they had no authority to stop anyone from entering the aircraft parking ramp. This was not a federal property. If there was an issue, we would need to call the county police.

I drove out to the ramp and didn’t see a problem so I parked the car and walked out into the crowd. This walk restored my faith in this country. These heartland folks were there to see the operations. There were farmers wearing bib overhauls. One drove 50 miles to get there saying, “I heard you flyboys were operating out of here. I wanted to see what you were doing.”

As I walked through the crowd talking with people, I met a family. The daughter wanted to be an Air Force pilot. We talked awhile. Finally, I said, “Go over by that car and wait for me.” I walked around talking with crowd members then went back to the car. I drove the family out onto the ramp to a B52 and took them up into the cockpit. I put the daughter into the pilot seat, mom in the copilot seat, and dad in the IP seat. I explained a few things about the plane and answered their questions. I wonder if she ever became a pilot.

Crowds of 300-400 arrived every day for our evening takeoffs. I asked any nonflying crew members to spend time in the crowd talking to folks. There were usually 20-50 folks there for our 2-3 am landings! I had the planes stop in front of the crowd to do the pre-takeoff last chance drive arounds. This was usually done on a taxiway near the end of the runway. I wanted to give a show to our supporting civilians!

I didn't advertise this in advance because I didn't want a huge crowd to show up but one day I held a spontaneous open house. I greeted the 300 watchers saying, "I'd like to invite you to walk out to a B52 and let you walk around it." I led the crowd out to the plane. I had non-flying crewmembers available to talk with the crowd and answer any questions they might have. The bomb bay doors hung down/open. There were dozens standing under the bomb bay. Most folks were just amazed at the size of the plane! After about an hour I announced that we needed to leave so the plane could be prepared for the night flight. We all walked off the ramp.

While deployed, there was a Wing Commander change back at Griffiss on 5 August. The Wing Vice Commander became the new Wing Commander.

The new Wing Commander visited Salina. There was only one air-conditioned room available anywhere in this deployment. It was a fairly large room in the old National Guard billeting building. When we first arrived in Salina, I told my staff that we would live in the same temperature conditions that the others lived in. That one air-conditioned room remained empty.

My new Wing Commander arrived and I gave him the air-conditioned room. That first night he was the only one sleeping in it. The next day he directed me to move into it with the senior deployed staff. We did because he's the boss.

The bombers flew 6 sorties each night, the tankers flew 2-3 sorties per day. There were operational work arounds. We were operating out of an uncontrolled airport. That means there was no control tower. Our B52s only had UHF radios, no VHF radios (civilian's frequencies). Departing and returning, the B52s would make required radio calls on UHF. We on the ground would then repeat their calls on the civilian frequency (VHF).

From our bare base, we flew 55 conventional B52 sorties into Twenty-Nine Palms and the Utah Test & Training Ranges. We also flew countless KC135 sorties. In the end, we learned about deploying aircraft to a bare base.

Bare base lessons learned here caused deployment plans to be modified. They were used during Desert Storm.

Diego Garcia

BUSY CUSTOMER, 4-17 Oct 88. My third trip to Diego Garcia occurred in 1988 when I was the 416BW Assistant Deputy Commander for Operations at Griffiss AFB, Rome, NY.

Diego Garcia is an atoll located within the Chagos Archipelago in the north-central portion of the Indian Ocean. It is approximately 1,200 miles south of the southern tip of India. The 6,720-acre atoll is horseshoe-shaped with a length of approximately 40 miles.

We flew three B52s to Guam. Our Alaskan KC10 tanker takeoff was delayed for two hours because of ice fog. We held off the Alaskan coast. As we burned fuel, we computed how long we could hold until we needed to divert to Fairchild AFB, Spokane, WA. The KC10 finally arrived, we refueled, then continued to Guam.

We then encountered severe head winds, over 100 knots. It looked like we may not get to Guam with the required fuel reserves. We called ahead to Guam and alerted the strip alert KC135 tanker that we may need them to come out and refuel us. Eventually the winds dropped off and we didn't need the tanker.

I was sitting in the instructor pilot seat behind and between the pilot and copilot seats. As we approached Guam the pilot (left seat) said, "I have a headache, would you make the approach and landing?" What pilot turns down a landing? Never turn down an offer like that! Yes, I made the approach and landing. Everyone went into crew rest in preparation for the flight to Diego Garcia.

The Guam Bomb Wing Deputy Commander for Operations and Deputy Commander for Maintenance Colonels called me to a meeting. I'm a Lt Colonel they are Colonels. They told me they couldn't supply the Guam maintenance personnel that the SAC operations plan tasked them to deploy with us to Diego Garcia. I said, no problem, and I could see a "we win" in their faces. But then, I asked to coordinate on the message I said they needed to send to the SAC/DO explaining why they couldn't support a SAC operational plan. I got their attention. They didn't want to up channel and explain their nonsupport, changed their story, and sent the required people.

From Guam, we deployed to Diego Garcia, 8-13 October. I flew ahead on the KC10. The best two B52s, maintenance wise, followed us to Diego Garcia.

I was on the ramp when they arrived. A gunner walked toward me with four .50 caliber rounds in his arms. He said they had been illuminated by airborne fighter radar off the Vietnam coast and he charged (loaded) the four .50 caliber guns. I told him to put the rounds out of sight. Diego Garcia was a British Protectorate Island and I didn't want to have to explain why we were walking around the ramp with live rounds.

The plan was for the two B52s to fly a low-level route across Somalia. The KC10 would fly out ahead, wait for them, and refuel them to maximum gross weight.

On the first fly day, the KC10 launched, then one B52. The other B52 experienced maintenance issues. It was fixed but too late to catch the other in time to make the scheduled low-level entrance time. I cancelled it.

The next launch day, the first plane airborne was the KC10 tanker. It experienced a malfunction, a "stick shaker" after takeoff meaning a possible stall situation? The KC10 dumped fuel to reduce weight to landing weight and landed. I cancelled the B52 flights to Somalia. I couldn't send B52s out without tanker support.

The KC10 fix was coming from Robins, AFB, GA. The KC10 couldn't be touched by military personnel or the warrantee would be voided. So, the man with the fix left Robins AFB. He was a Filipino who got "lost" for a couple days in the Philippines.

Days went by waiting for the KC10 fixer. The flight crews were enjoying the sun and salt water. I held a short staff meeting to discuss and update the staff on the situation. I asked the maintenance chief how long it would take to put a B52 in the air for a local flight. He said about an hour. I said we would be there in an hour and half. I looked over at Harry and asked him, "Do you want to fly?" Harry said, "Hell yes!" There were three non-pilot Lieutenants on the deployment at the meeting: one security police, a maintenance officer, and one from supply. I told them they were getting a B52 orientation ride!

After takeoff we climbed to 20,000 feet. I cycled the three lieutenants through the copilot seat and I gave each the opportunity to manually fly the B52. Harry and I then made a bunch of touch and go landings. We then flew north of the island. I called the Navy Harbor Master and asked him for permission to fly through the harbor at a low altitude. He approved, and added, "Just don't hit any of my prepositioned supply ships anchored in the harbor." I increased the airspeed to maximum and descended to just below 400 feet. We blew through the harbor entrance. Near mid harbor we pulled the nose up into a high angle/rate climb trading airspeed for altitude. We then returned for a full stop landing. That was my one and only joy ride in a B52! I still wonder what the three 2nd Lieutenant's thoughts were about their flight.

So, we waited for a few days for the KC10 fix it man. Finally, he arrived on a C141. I watched as the C141 taxied up next to the KC10. The fix it man went up into the KC10 carrying a small box. He pulled out the malfunctioning avionics box and shoved in the new box. It took days to get there but only about 30 seconds to fix the problem!

We were on a deployed time line and had to redeploy back to the states. We left after flying only one planned flight out of Diego Garcia. That is, except for the Harry and my B52 joy ride! We flew back to Griffiss.

Occasionally an inadvertent Klaxon would occur for various reasons. Crews responded and started engines, and they began asking for the Emergency Action Message (EAM), "Control say message." Of course, the inadvertent klaxon was a surprise to the controllers and they didn't have an EAM from Headquarters (HQ). They checked with HQ, then created inadvertent klaxon EAM that they read to the crews. The message told them it was inadvertent, to shut down, and return the aircraft to alert status.

While I was the DO at Griffiss we experienced one inadvertent klaxon. I got a call from the command post and responded to find the Wing Commander listening, repeatedly, to command post tapes of the event.

He and I talked, he was convinced that the command post controllers had committed a mistake. I wasn't convinced they screwed up. I talked with the command post controllers about the Klaxon event. While I didn't know why the Klaxons blew, I believed it wasn't a controller team error.

Keep in mind, I had extensive command and control experience. I been in the Fairchild Wing command post for 3 years. Then in the CINCPAC, Airborne Command Post.

I mention this because I knew the command and control world.

I lost the argument when the Wing Commander directed me to decertify the command post controller team. That meant they were removed from the console and replaced. They would need to attend some sort of retraining. I argued to no avail that it wasn't their fault. Being a new Wing Commander, I think he believed he needed to show HQSAC he acted. He destroyed their military careers.

SAC ORI

I experienced a Strategic Air Command (SAC) Operational Readiness Inspections (ORI) as a crewmember, a command post controller, and finally as a SAC Bomb Wing DO. An ORI was a really big deal, for everyone! The Wing was evaluated by the SAC/IG to see if it could accomplish its nuclear war mission.

The SAC/IG arrived in March 1989. The 60 inspectors dispersed throughout the Wing and across the Base. An engine start exercise initiated the evaluation. What followed was a complete recall of all wing personnel. Units were graded on the speed of which personnel assembled.

Historically, after crewmembers responded for the recall, they were released until needed to preflight a generating alert aircraft. SAC had just broadcast a new concept called, "Survival Launch." This concept called for a Wing to have the ability to launch on warning generating and non-alert aircraft, to survive them.

I had developed a new Wing plan. Instead of releasing the crews after recall, I keep them in the alert facility. They were assigned an aircraft and were responsible for a Survival Launch of their assigned plane.

An aircraft generation began. Every B52 was loaded with a nuclear war load and put on alert. Tankers were loaded and configured for alert. The Wing had a generation schedule that depicted how many aircraft were generated to alert over time.

Generation complete and graded, nuclear weapons were then downloaded and aircraft prepared to fly a simulated nuclear war sortie. All crews went into crew rest.

The IG tested the crews with written tests about their responses to various Emergency Action Messages (EAMs) and aircraft performance questions. Two KC135 Tanker

Training Flight instructor pilots appeared in my office. They said that tanker crewmembers were failing the IG performance test because a particular question. They maintained the IG question was wrong. They showed the question to me and explained why it was wrong it to me. I said, "Have a seat."

Recall that my last assignment was at 1CEVG and I knew the KC135 tanker evaluators. I called them and asked if they would look at the IG performance question and the answers to see if it was correct. They did. The IG question was wrong.

I went down to the IG working office and explained to them that the 1CEVG tanker evaluators reviewed their performance question and it was wrong. The IG dropped the question. The tanker crews passed! It's good to have friends!

The ORI occurred over several days. I was in my office talking with my chief Electronic Warfare Officer. I had been up straight for over 2 days. He tells me that as I was sitting in my chair, I fell asleep in mid-sentence. He quietly left my office.

Fly day arrived. These flights needed to be perfect. 15 bombers needed to takeoff, one every 15 minutes. I'm the DO orchestrating the launch. I was on the ramp in my car with my car UHF radio and hand carried brick (radio). Sortie 5 was having a maintenance issue. I pulled up in front of sortie 6 and called over UHF, "Are you ready to fly sortie 5 timing?" "Yes" was the answer. I respond, "You are now sortie 5." They taxied out. The old sortie 5 was fixed and flew as sortie 6. All the planes launched.

A B52 taxied out. As it was making a left turn, I saw fuel pouring out from the right-wing tip. Concerned I drove up behind it and called out for it on the UHF radio to stop, it did. I then said, "The fuel is from the over filled bowser tank. Once the wing was airborne and lifted this fuel would drain back into the main wing tanks." I said, "I clear you for flight, contact Tower for takeoff clearance."

One airborne B52 had to air abort because of maintenance problem and was returning. It was a peacetime maintenance issue. The aircraft crew chief heard his aircraft was coming back and tears welled up in his eyes. This proud young man wanted his aircraft to succeed. I explained to him that in a nuclear war his plane would continue to its target with this problem, according to peace time rules, it needed to return.

The IG completed the 10-day ORI. We did extremely well. Seven of my eight divisions received an excellent rating. The IG report commented that this was the best aircraft generation the team had seen. The IG report also said our ability to deliver our weapons on target was rated 98%, the best in SAC for the ORI cycle. Both squadrons, bomber and tanker, were rated Outstanding.

I have lost friends during peacetime training, and in war, friends died. My most proud personal accomplishment at Griffiss was that I didn't lose a single crewmember.

My focus during my Griffiss time was realistic war fighting. A couple short years later these crews were flying combat sorties during Desert Storm.

When I was selected for Colonel, the board also selected me to attend the National War College located at Fort Lesley J. McNair, Washington, DC. I left Griffiss still a Lieutenant Colonel and headed for Washington, DC.

Chapter 8 - National War College Student

There are four Senior Service Schools, War Colleges: Army, Navy, Air Force, and the National War College.

The National War College (NWC) is located at Fort Lesley J. McNair, in Wash, DC. It is a 10-month course in national security strategy that prepares future military and civilian leaders for high-level policy, command, and staff responsibility. There were 170 students from Military and Government agencies. The 170 students were organized into 10 committees. I was in committee 7. We were the class of 1990. Shortly after arriving I pinned on the rank of Colonel.

I rented a 5th floor apartment in DC on 4th street. It was a great location just a couple blocks outside of Fort McNair. My furniture arrived and was brought up the elevator to the apartment. That is, except for a couch. I was told it was too long to fit in the elevator. It would be put in the basement. The movers would come back in a few days and hoist it up on the outside of the building, onto the balcony, and then into the apartment. When you move as many times as I have, you have moving stories.

A few days later they came back to lift the couch up to the apartment. I joined one mover on my 5th floor balcony. He threw a single yellow strap down to the four men on the ground below. They looped the strap around the center of the couch. Curious, I asked, "I see four men standing down on the ground and only you up here. You got this?" He said, "Yes, I've got it." This one man starts to pull the couch up.

He pulled it up to about the 4th floor, then the single center strap loop slipped to one end of the couch and the horizontal couch went vertical. The moving man and I glanced at each other briefly. Then strap slipped off the couch end. I watched the couch grow smaller as it fell. It landed on the end with a crunch and fell over. The four guys on the ground converge on it.

The man next to me yells down, "Is it ok?" "Yes" was the response. They now made two strap loops around the couch, one at each end. The man next to me starts to pull it up again. I asked, "If the couch is damaged, will you come back and lower it back to the ground for repair, then back up here?" He says yes. He pulls the couch up, over the banister, and into the apartment. One couch arm was broken. We placed the broken arm end against the wall. It was a very old couch and not worth fixing. I lived there 10 months. The moving out story was even more interesting.

The National War College curriculum was intense! I graduated from civilian colleges with a Bachelor of Science Degree, and then earned a Master's Degree from Pepperdine. These military 10-month schools were total immersion.

Over the 10 months we listened to 139 distinguished speakers. There was a broad core curriculum and electives covering the art of war, strategy at the national level, US security interests, and national security policy.

The purpose of the school was to learn about how the government agencies worked and to develop better understanding of all the military services. We visited the White House, the Capital, the State Department, NSA, and FBI headquarters. There were trips to Nellis AFB, Norfolk, Fort Benning, and Camp Lejeune. We studied the Battle at Gettysburg then travelled there and walked Pickets last charge.

I lived on 4th street which was walking distance from the Capital Mall. I watched money being printed at the US Treasury and toured all the sites and Museums around the Mall.

As part of the curriculum, the class was split up and went on 16 separate international trips. Throughout the month's students studied the countries they would travel to. I studied North Africa: Morocco, Tunisia, Algeria, and Egypt.

On the trip, we visited our US Embassy in each country and received a country briefing. We then met with senior military and civilian leaders in all the countries. The schedule was intense. By the end of the trip I was ready for it to be over.

Our last country was Egypt. We were there on one of their religious holidays so nothing was on the schedule. I had been to Luxor before so I suggested that we airline down to Luxor and tour Karnak's Temple, Valley of the Kings, and King Tuts Tomb.

Somehow, we were seated in the first row of the airliner just next to the entry door. All the seats in the airliner were the same. There was no first-class divider wall between rows. The Captain came in and we began to talk. He gets excited when he learns we are military and invites us into the cockpit. I'm first in and sat on the jump seat behind the captain's seat. The others stand behind. The crew included the captain, a copilot, and an engineer.

To demonstrate how they were prepared to live in the harsh desert Egyptian Rangers Kill, Clean, and Eat Snakes....Raw

The captain told us he had been a fighter pilot and said he had been shot down several times by the Israelis during the various wars. Then he says, "It's time to start engines. One of you can stay." I'm in the jump seat so I wave bye-bye to the others. He talks with me through engine start, taxi, takeoff, and the climb. After level off, he said, "Go back and have lunch, then come back for the descent and landing."

So, I came out of the cockpit door and am standing in front of my group. I look back and see rows of people all looking at me. I decide to screw with people and said to my group, "He let me make the takeoff. It flies like a KC135. He told me to have lunch then go back and make the landing." The five rows behind my group immediately erupted into intense chatter, in French, I think.

I had lunch then went back into the cockpit. We started down. He said, "Do you know how we find Luxor?" I respond, "I have no idea." He says, "See that bend in the river? We descend to 4,000 feet over it and turn to heading 270 and fly for 4 minutes."

I see the Luxor airport and we are soon descending toward the runway. He holds up the paper approach chart and points to a symbol on the chart and asked, "What is that?" I say, "It's a VASI (a glideslope light system that guides pilots to the runway.) He points to the runway and

Rick the Camel Jockey

asks, "Where is it?" I responded, "I can't see it." He said, "There isn't one."

In Luxor, we saw all the sites, an amazing place. That night my flight back to Cairo wasn't as much fun as the ride to Luxor. During the international trip we all eventually experienced extreme digestive bug issues. My turn occurred on the flight back to Cairo. It was really nasty.

Did I mention that I rode in an Egyptian airliner cockpit! I still can't believe it!

Lots in the world changed during the 10 months at NWC. The biggest event was the fall of the Berlin wall and the collapse of the Soviet Union. These events would have immense impact on my next assignment and the military.

My assignment after the NWC was back to the Strategic Air Command as Chief of the Mobile Command and Control Division, at Offutt AFB, Omaha, NE. With a $127 million budget, I would oversee the SAC Airborne Command and Control System.

This assignment made sense to me. I knew the nuclear war command and control system. I had been a B52 crewmember at Beale, a command post officer controller then a training officer at Fairchild, and finally, I was a battle staff operations officer on the CINCPAC ABNCP in Hawaii.

When I received the assignment, I coordinated the timing for travel, leave (vacation), and arrival dates with the assignments section. The plan was to go to Omaha, search for a house, then go on leave and spend some time in Colorado. I had my orders in hand.

I was in my committee 7 room when the phone rang. It was my soon to be command post boss. He is a one-star general who tells me he needs me there immediately since there is lots to do. I respond that I assumed he was involved in the assignment coordination timing and my plans were finalized. Finally, I said, "You won't have my security clearance before my planned arrival time so I wouldn't be able to see classified." We argued several minutes, finally he gives up and said, "Ok, but, can you at least stop in the office for a few minutes over coffee so we can talk?" "Yes," was my answer.

Conversation over, I hung up the phone and turned around to find two marines sitting in my committee room staring at me in disbelief. One said to the other, "Can you believe it? He just told his new boss to F…off. We would have just saluted and showed up."

I was in Washington DC for 10 months and had travelled all over the city. I knew a number of folks who were assigned in the Pentagon. I avoided any contact with anyone there until I had an assignment. I had visited the Pentagon a number of times through the years and didn't want anything to do with it.

Assignment in hand, I was now safe to visit friends in the Pentagon. I went down into the basement level to the Tactics Division located next to the purple water fountain. I had a great visit with a friend who pointed to an empty chair saying, "You should have let us know you were in town that could have been your chair." I respond with a sad, "Darn."

Another moving story. I had a move out date and had travel plans. I needed to drive away by noon. The truck driver and helper showed up to pack and load my stuff. I led them from the apartment on the move out route through the building. I led them down the elevator to the basement and walk through the basement to stairs up to the loading dock.

They are ahead of me as we walked down the sidewalk alongside of the building, I hear the driver say to the helper, "I'm out of here." And they walk off and depart without another word!

I called the moving company to tell them what happened and that I needed to leave by noon. Thankfully, five moving company packers responded to move me. One had no shoes on! He had gone to the office to pick up a paycheck and agreed to come along and help pack me out.

I left on time. I left Washington, DC, or tried to. The traffic pattern around the area was a bit confusing. I ended up driving around the Pentagon five times before we (Renee was following) were finally on our way to Offutt, NE.

Chapter 9 - Mobile Command and Control Division

When I received the assignment, I called ahead to talk with the man I would replace. He said the operation ran smoothly, like it was on auto-pilot, and it had not changed since 1961. Boy was he wrong. While I attended the National War College, the Soviet Union collapsed ending the Cold War.

"Looking Glass" was an EC135 airborne command post. There was a command post airborne continually around the clock since 1961. They flew 8-hour shifts. The battle staff and communications it carried were capable of directing our nuclear forces.

It was a backup in the event the Soviets destroyed the SAC underground command post at Offutt. Also, there was a 27 plane Post Attack Command and Control System (PACCS). They had the capability to launch our nuclear missiles from the air. There was also a ground mobile command and control system that could deploy. The National Emergency Airborne Command Post (NECAP) was the war plane that would carry the president. One always follows him around.

After I arrived, our command and control structure would undergo massive changes. Looking Glass was put onto ground alert, then was shut down, and later SAC stood down and was replaced by US Strategic Command (USSTRATCOM).

When an order came down to put the Glass on ground alert, there was a mad planning scramble on how to setup. The plane needed a guarded parking location, aircrews and battle staff needed sleeping quarters and food, and finally enough vehicles to move all the personnel rapidly to the alert plane. We got it done. The Glass was on ground alert.

Iraq invaded Kuwait, then threatened Saudi Arabia. The coalition military buildup ended when Desert Storm began. The Senior Battle Staff went into 24-hour operation for over 3 months. We were on 12-hour shifts which meant being there 13-16 hours. I was a Chief of the Battle Staff. Each functional area had representatives on the staff. We coordinated and worked issues for the deployment and bed down of our B52s and KC135s. Each morning I briefed the SAC senior general officers on the buildup status and any occurring issues.

Iraq was firing ground to ground (SCUD) missiles into Saudi Arabia. I was familiar with how the Cold War missile launch warning system worked. We called the North American Aerospace Defense Command (NORAD) and asked if they would give SCUD missile launch warnings. Their initial response was no, not their job, they were a nuclear war operation. Our general talked to their general. Soon they were on board. When a SCUD was launched, we and the theater received launch warnings.

During my time as Chief of the Mobile Command and Control Division, I was involved in the Top-Secret world and a number of Special Access Code Word programs. They were strict "need to know." One day during Desert Storm, my boss called me in to his office and asked me about some KC10 flights. I told him he didn't have a "need to know." Boy, was he pissed!

I lived in the highly classified world. Ever been drug tested? I was, many times. Usually, I was handed a letter telling me to respond within 2 hours to a bathroom where a man would watch me wiz into a bottle.

One day I was handed a letter that directed me to report to a room for a lie detector testing session (1 to 4:30 pm). In 1990, an Army team was sent to Offutt to give lie detector tests to Special Access Program, Code Word, officers. I was given a show time and a room number. A female Army Sergeant met me. She discussed, in depth, the 5 specific questions she would ask me. She spent time explaining the meaning of each word. She told me that nobody could beat the lie detector and explained in detail how it worked. I was then wired up. She and the machine were behind me, I faced the wall.

She asked the five questions in turn and I'd answer, No, to each.

1. Have you ever engaged in espionage with an agent of a foreign government against the USA?
2. Do you know of anyone who has engaged in espionage with a foreign agent against the USA?
3. Have you ever stolen supplies from the government?
4. Have you ever falsified official documents?
5. Do you ever leave classified documents out unsecured?

I could hear the machine pen scratch in response to each of my answers. After awhile, she started to ask me if I was sure of my answers. She said the lie detector was recording something strange, something was not quite right. Am I sure? Am I sure? Am I thinking about something else? What am I thinking about? After awhile she said she needed to leave the room and talk to a supervisor about my "strange" results.

Me, I had already reverted to my 1970 survival school POW training (20 years earlier) and I don't believe anything I see, hear, or am told. My guess was the whole show was to put mental pressure on me to see if I would confess to some wrong doing that I might be engaged in.

She would return and throw out a question, then start again. This happened a couple more times. She comes back and says she will throw out a question. And we begin again. Then suddenly at 4:30 pm she declares the lie detector test over.

With the EC135 airborne alert mission going away, we looked for other uses for the aircraft. They carried amazing communications capabilities. We started working with the Army Special Forces using the airframe as a standoff communications relay node. One of our EC135s slid off the end of the runway landing at Pope AFB. No one was hurt but the plane was broken.

My one-star boss had been a F4 fighter pilot. He asked me how this could happen. I guessed, "Landing on a wet runway with a tailwind without checking the performance book to see what the required stopping distance would be." Sure enough, that's what happed.

SAC would stand down to be replaced by US Strategic Command (USSTRATCOM) on 1 June 1992. There were 20 outstanding officers in my Division. Their entire professional world would simply disappear! To a person, they came to me for guidance. I had nothing. I had always taken care of my people. But here, I was powerless.

Change caused turbulent times. The Looking Glass operation had been ongoing since 1961. Now it would end. A reduced command post battle staff function would be transferred to the Navy "Take Charge and Move Out" (TACAMO) E6 aircraft at a different base.

There would be a massive reduction in manning needed at Offutt when SAC closed and STRATCOM stood up. All unneeded folks, including me, were assigned into Detachment 2. It was a holding place for the unneeded personnel until retired, or reassigned.

I had joined the Offutt Aero Club when I first arrived. It had an impressive fleet of small aircraft. While in Detachment 2, I used my GI Bill benefits to earn two FAA flight ratings: Airplane Transport Pilot (ATP) and Certified Flight Instructor (CFI).

With the Cold War and Desert Storm over the military force size drawdown continued. When I entered the service in 1969 there were 862,353 in the Air Force. By 1990 (535,233). In 1991 (510,432). The numbers reduced every year to 2017 (321,125).

This was at a time when military forces were returning from Desert Storm while a massive Reduction in Force (RIF) was occurring. Our war fighters were returning from war then getting released from the military.

A Selective Early Retirement Board (SERB) was scheduled to down size the number of Colonels. We were given a choice. Apply for retirement before the board meets and retire one year out, or, meet the board and if selected, retire in three months. The board selected 30%. Fortunately, I was not selected to retire. I survived three (30%) SERB boards.

I finally received an assignment. I would be joining the Air War College faculty at Maxwell AFB, Montgomery, AL. I was a bit apprehensive about this assignment because being on the faculty would be a totally new experience for me. During my drive to Maxwell I recall dreading it. Every other assignment I have had, I looked forward to.

I still had GI Bill benefit money to use. I earned my Master's Degree from Pepperdine in 1976. I called the Maxwell Education office and asked about my education options. The University of Alabama offered a Doctor of Public Administration (DPA) degree. It could be completed in 36 months. The classes were given on the base. The first class would be on 16 September, I would arrive on the 10 September. I applied.

Chapter 10 - Air War College Faculty

I arrived at Maxwell AFB and joined the faculty on 10 September, 1992. I had been a student at National War College in Washington, DC. The Air War College was the Air Force equivalent to the National War College. We lived in temporary base quarters for 3 months before a house on Maxwell became available.

There were 250 total Colonels in the student class, including 38 international students. The 10-month long course began in August, finished in May.

The Air War College curriculum stressed the application of air and space power in joint/combined operations through six themes of leadership, doctrine, strategy, technology, political-military integration, and joint warfighting. Usually, there was an hour and half stage lecture about the topic being studied. A question and answer period followed. Finally, the students moved from the auditorium to the seminar rooms where the lecture and subject issues were discussed. In the afternoon, students went to elective classes.

I was assigned as Director, Leadership Management Studies. The Air Force had embraced the Total Quality Management (TQM) concepts. We called it Quality Air Force (QAF). I was tasked as course Director for the first ever course devoted to Senior Leadership of QAF. I developed the curriculum that resulted in materials for 18 educational classes. I was known as "Mister Quality" at the Air War College.

I was also Mister Myers Briggs Type Indicator (MBTI). I flew an aero club plane to Gainesville, FL, where I attended MBTI training. The MBTI instrument is not a test but asks a lot of questions. There are no right or wrong answers. The pattern of answers defines a person's personality and how that person processed information. I'm an Introverted, Sensing, and Thinking, with Judgment (ISTJ).

MBTI question instrument was sent out to all incoming students and their families. Their responses were processed. The entire student class and their families attended a lecture by Dr. Otto Kroeger who explained what their results meant. As he talked explaining what the results meant, I could see family members pointing at each other as they suddenly realized their thought process.

Part of the curriculum included the Regional Studies Analysis (RSA) program. Areas of the world were studied (Usually 3-4 countries for each travel group), then a 12-day travelling seminar was made to the countries. Travel members visited the US Embassy, then met with country senior government, and military leadership.

There were 14 trips scheduled. I had experience in Middle East so was assigned to the Middle East field studies trip mid-way through the preparation phase. We studied then traveled to: Riyadh & Dhahran, Saudi Arabia; Tel Aviv, Israel; Cairo, Egypt; and Lajes, Azores. We flew on an Air Force KC135 tanker.

We arrived in Riyadh in the evening, put our bags into our quarters, and were driven to a gold Souk to shop. I got off the bus. Walking through the Souk look down to discover my wedding ring missing! Talk about a case of big eyes! I traced the route back to the bus scanning the ground for the ring, no ring. Back at quarters I crawled around my room looking for the ring, no ring. I searched around my seat in the plane, no ring. I called back to Maxwell and had a friend look on the ramp where we loaded the KC135, no ring.

From Israel, I called home and chatted with Renee but I didn't mention the lost ring. I was hoping to hear that I had left it there, no ring.

We return after the 12-day trip. I took my bags into the house. I had carried a paper bag on the trip that contained snacks like Cup-of-Noodle type treats. The paper bag sat on the floor next to my seat where all I had to do was reach down to pull out a snack. There was one snack left and as I pulled it out, I see my ring on the bottom of the bag! I'm sure glad I didn't finish the last snack on the trip and perhaps thrown the bag out. I had my ring size reduced!

A year later I led this same trip with the next class. I didn't lose my ring this time. These trips were grueling. Everyone is worn out and was always glad to be home.

The Cold War ended when the Soviet Union collapsed in 1991. Russia sent the first officer to attend the Air War College in 1992. He graduated in the class of 1993. The Russians made the mistake of letting his family travel to the US to attend his graduation. After graduation he defected. He became an instructor at a university in Dothan, AL. I don't know if Russia ever sent another officer to attend the AWC.

Fire Ants

I've given CPR once and received CPR once. In 1984, I was in the B52D Squadron on Anderson AFB, Guam. I was Chief of Training Flight. After a training sortie, I was debriefing the pilot and copilot. Bert sat down at the table with us and listened in on our discussion. As I debriefed the two, I looked over at Bert, he looked really bad, and so I asked. "Bert, are you ok?" He says, "No, not really" as he passes out, falls over, and drops to the floor. I check him. He is not breathing so I begin CPR. I yelled out, "Call for an ambulance." Bert started to breathe again and came around as the ambulance arrived. They hauled him to the base hospital emergency room. He survived, but never flew again.

Years later, it was my turn. I was a Seminar Director at Air War College at Maxwell Air Force Base. We were going to have a seminar party at our house on Maxwell on Saturday night. Late morning, I was mowing the lawn. Almost finished, I decided to cut taller grass next to the garage that I usually didn't cut. I was wearing shorts and low-cut sneakers. I went into the taller grass to look for any sticks. As I stood there, I started getting stung on my ankles. Looking down I see about 100 fire ants on my legs and quickly brushed them off.

Cutting the taller grass idea was cancelled and I pushed the mower into the garage. As I walked toward the house, I became weak and squatted down putting one hand on the ground for support. Renee happened to be near. I said, "I don't feel very good." She looked down at me and immediately said, "Come on, we're going to the hospital." Later she told me that my eyes and mouth were already beginning to swell shut.

I walked behind her up the five stair steps, intending to follow her into the kitchen where she would get the car keys. I got halfway up, passed out, and stopped breathing. Renee came out, looked at me realizing I wasn't breathing, and gave me two rescue breaths. A doctor lived next door. She quickly ran over, banged on his door, he wasn't home, and ran back to me gave me 2 more breaths. She ran into the house to get the mobile phone, (no cell phones back then) and dialed 911.

The phone was a city line from the local phone company, so when she dialed 911, she was connected to the City of Montgomery dispatch center, not the Base. She quickly told them our emergency. She asked them to call Maxwell Air Force Base Security police and have an ambulance sent to our address immediately. She came back to me and continued rescue breathing. My heart did not stop so I didn't require compressions. The street we lived on was a large circle. The base hospital was only about three blocks from us. She heard the ambulance siren and is relieved, however, they are taking the long way around the circle.

I'm coming around as they arrive. I was loaded into the ambulance and we make the short trip to the base hospital emergency room. The ER Doctor wasn't sure how to treat me. He said to Renee, "I think this is a vasovagal reaction (blood pressure gets too low and you pass out)." Renee told him "It is an anaphylactic reaction to fire ants. Give him some Epinephrine and some Benadryl." His response, "How much?" She tells him, since he is a big guy start with 0.3mg of epinephrine (that is what is in an Epi-Pen) and 50mg of Bendaryl.

The nurses were staring at Renee and shaking their heads in disbelief. Renee looks at them and screams, "Do you want me to start that IV?" They said no and got busy! Renee was pissed! As per protocol in an emergency of one of the Air Force members on the base, my Commander (a general), showed up. Renee read him the riot act of having incompetent doctors taking care of their most valuable resources, meanwhile poking her finger toward his chest, not quite making contact. I wonder how many times he had that happen from a spouse. In any event, I recovered, we returned to the house.

We still had the seminar party that night. I spent most of it propped up on the couch. BTW, that was my one and only ambulance ride, 3 blocks long. I carried an Epi-Pen for a couple years, then stopped. Since then, I've been stung by fire ants many times without a reaction.

Renee's parents visit

Renee's parents stopped in for a visit on their way north to upper New York. They were snow birds, traveling between Florida and New York. We had a great time. While they were there, our wedding ring inspections became

Cessna T41, 4 seat, Single Engine

due. The nearest Kay's Jeweler was in Dothan, AL. I decided to take the family for an airplane flight south to Dothan. I rented a T41 from the Maxwell Aero Club. The 4 seat T41 was a supped-up Cessna 172, with a bigger engine and climb propeller.

Dad sat in the front seat with me and the ladies just behind us. There was a solid cloud overcast so I filed an Instrument Flight Rules flight plan. As we flew along, I explained what I was doing and why. We were in the clouds the entire flight so there was no sightseeing. The flight was smooth. When we arrived at the Dothan airport, I flew a non-precision instrument approach, broke out of the cloud base, and made a landing.

We drove into the city, had our wedding rings inspected, and then enjoyed a nice lunch.

The weather was still solid clouds, so I filed another Instrument Flight Rules flight plan for the return trip. It was another smooth flight back to Maxwell. Because of the reported overcast I planned to fly an ILS (Instrument Landing System) instrument approach. It is a precision approach that provides both line up (Lateral) guidance and a glide slope that would guide us down to 200 feet over the ground. To fly it I would need to fly about 8 miles north past the runway, turn around, and fly the approach back.

Renee and Parents Survived a Flight with me

As we were passing Maxwell we broke out of the clouds and I saw the runway down and to my left. Now there was no need to fly the 8 miles out, then turn around and intercept the ILS and fly 8 miles back. I called the control tower and requested a visual approach. Request approved, I pulled the power back, lowered the nose, and began a descending left turn toward the runway.

As I taxied down the taxiway toward the aero club I'm thinking, "What a perfect day, flew Renee's folks for the first time, and showed them the magic of instrument flight."

The aviation environment can change in a second. As we taxied up in front of the aero club, about 10 members come out the door. A number of them are pointing at our plane. All are talking. My first thought was, "That's not good." I shutdown the engine and we got out. The entire left side and bottom of the plane was covered with engine oil! No oil showed on the oil dip stick when I checked the oil level!

I told Renee's folks that it's better to have an emergency and not know about it until the flight is over. The oil pump ruptured so engine oil was pumping out as we flew along.

I'm a lucky man. If I hadn't seen the runway and asked for a visual approach, or, if I had decided to show them how an ILS worked, we would have had to fly out, then back over 16 miles. We possibly would have lost the engine and had to make engine out landing.

Chapter 11 - AFROTC Assignment

I graduated college on 7 June, 1969. I completed Air Force Reserve Officer Training Corps and was commissioned a 2nd Lieutenant in United States Air Force. Fast forward 24 years.

In 1993, I applied for a ROTC duty assignment as a Detachment Commander. I applied with a letter, resume, and official pictures. There are about 144 colleges across the country with ROTC programs. Assignments were for 3 years.

That year, there were 70 applicants for the available ROTC openings. A selection board assembled and reviewed officer records. Based on the board record reviews, applicants were racked and stacked listing them 1 through 70.

My record must have competed well compared to the others because a few weeks after applying my office phone rings. The ROTC assignment voice asks, "Do you want to go to the University Southern California (USC)?" Now I've talked to the Air Force commander at USC. He told me it was in a rough part of town in Los Angeles and that he had kept a baseball bat in his office for office defense. I asked the ROTC voice, "If I turn the offer down will I be dropped from further consideration?" He says, "No." I turned down the USC offer.

The next week the ROTC voice calls again with a position at the University in Portland, Oregon. This sounded really interesting but, and I'm not sure why, I say no.

A week later I received another call. My third offer was to the University of Texas at Austin. I had never been to Austin but had always heard great things about the place.

It was Friday afternoon so I asked the voice, “Can I have the weekend to think about it.” He said, “Yes.” I went out into the faculty hallway and yelled that I could go to Austin, should I? All the yells that came back were, “Hell yes, go there!” I went to the base library and picked up a bunch of literature about Texas and Austin. Renee and I looked over the literature on the weekend and we decided, yes.

On Monday, I called the ROTC voice and told him, “I’ll take the Austin offer.” The voice said, “You will need go to the university and interview with the University ROTC committee. Then if they approved you, the moving date is about 4 months out.”

We rented a twin-engine Cessna 310 (6 seats) from the Maxwell AFB Aero Club and flew to Austin. As Renee and I walked through the Aero Club, I said, “It’s about 3 hours to Austin so you better use the bathroom now.” She did. Just minutes later we were airborne and I leveled off at 12,000 feet. The plane does not have autopilot so I’m hand flying it. I get a couple tugs on my shoulder. Renee needed to go to the bathroom, again, already! I can’t believe it!

Agitated, I grumpily said, “Do you want me to land this plane?” With that, she hops over the seatback and disappears in the back for a few minutes. She climbs back over the seatback into the front seat. Curious, I ask, “So what went down?” She answers, “I used a zip lock bag.” We flew west to Austin.

I met with the university ROTC committee on Friday afternoon. It consisted of 6 university professors. While my academic history and grades were totally under whelming, I could talk a good story. Actually, they were impressed with my breath of my experiences and military record. The committee signed off on me.

Renee and I then went house hunting on Saturday/Sunday. The realtor took us around to a number of housing areas, but no luck. Then on Sunday afternoon we were at a development model home. I told the representative, “We are looking for a one-story house. And since we intend to sell after the 3-year assignment wanted nothing weird or unusual.” He said, “I don’t have any single-story houses built yet but follow me.” We walked about two blocks. Renee and I were standing on the sidewalk looking at a concrete slab with various pipes sticking up. The rep shows us the paper floor plan and said, “This house will be built here.” We look at the plan, at each other, and I said, "We'll take it." We went back the model home, took about 30 minutes to sign papers, pick colors, options, and then left for the airport and flew back to Montgomery. I wonder how many people have ever bought a concrete slab (house) in 30 minutes.

A couple months later, Renee travelled to Austin to interview for a job. She visited the house construction. A couple of things were wrong. There was no whirlpool tub in the master bathroom and no island in the middle of the kitchen. The tub issue was fixed. The island couldn't be put in without jack hammering in the concrete floor to put in an electric line for power on the island. So instead of the kitchen island, Renee took a number of other upgrades to the house. She asked why the floor molding was already in but not the flooring? She asked. "What if the molding doesn't meet the flooring?" They said, "We will caulk it."

After I received the assignment to the University of Texas at Austin, I heard about an awesome program. There was a number of ROTC detachments across the country with a Civil Air Patrol/ROTC a joint agreement. ROTC cadets paid $15 to join CAP. They then received 8 hours of flying, 4 hours in the front seat and 4 hours in the back seat. Local CAP Cessna 172/182 aircraft were used.

The ROTC Headquarters was located on Maxwell AFB. It was a short walk from my base house. I walked over to the HQ and located the Captain who ran the CAP/ROTC program. He had a map of the US on the wall behind his desk. There were pins stuck at the locations across the country where the program was occurring. I asked him a few questions about the program, then went behind his desk and pulled open a drawer in his desk. I picked up a pin and stuck it on Austin. I said, "There, you now have a new program at UT Austin." He smiled with a "Yes Sir!"

We moved to Round Rock, TX. The house wasn't completed yet so we lived at the billeting (Small apartment on base) at the Bergstrom AFB for a week. We were probably one of the last folks to stay on Bergstrom. The base closed and became the Austin Bergstrom International Airport. (BTW, we bought a house in 1994 believing we would own it for only three years. Well it's 2018, so 24 years later we are still in it!)

A week after we moved in, we were in the house sleeping and heard a bunch of noise coming from the roof! We went out the front door to see what was happening. There were 5 ROTC cadets on our roof and others on the ground! A good roof stomp is a great way to welcome the new Detachment Commander and wife! They jumped off the roof onto our newly laid sod. They left foot prints in the new sod. Renee and I invited everyone into the house where we meet our new friends. That was the start of a great 3-year assignment!

Shortly after Renee and I arrived we heard about a great restaurant called The Oasis. One Sunday we went exploring and drove out to it for lunch. It is indeed a special place with a beautiful view. The large restaurant sits on very high ground overlooking Lake Travis to the west. There is a series of wooden terraced decks down the hillside. Their food specialty was Tex-Mex.

We sat out on one of the decks, enjoying the day, and having a great lunch. The view was great, I saw off in the distance what I believed to be beach area. There appeared to be a parking lot near the road, with a tree barrier between the beach area and the parking lot. I'm thinking it's something like a state park.

I point it out to Renee and suggest we explore it and walk the beach. After lunch we drove toward it. We found it, parked, and proceeded to walk a trail through the forest.

We came across two guys walking toward us. They didn't have any clothes on! We were slightly shocked. We accidently discovered Hippy Hollow, a nudist park. We were the only folks with clothes on. We didn't stay very long. Renee jokingly said she should have handed out business cards because some folks needed fixing– she is an OR Nurse.

Later, I told the cadets about our accidental discovering of Hippy Hollow. They all had a big laugh.

There are four ways to become an officer in the Air Force. All are extremely competitive. They are: be selected to attend and graduate from the Air Force Academy, Officer Training School (OTS), get a direct commission, or complete in Air Force Reserve Officer Training Corps (AFROTC) at one of 144 universities.

I was Professor of Aerospace Studies at AFROTC Detachment 825. In 1994, there were 48,000 students at The

University of Texas at Austin. Organizationally, we fell under the Dean of Liberal Arts School. I received excellent support from him. There were three other officers, two administrative NCO's, and a Secretary. The Army and Navy programs were in the same building.

At the beginning of the school year there were about 120 cadets in the corps. That number decreased to about 80 throughout the year. I taught the sophomore classes on Monday and Wednesday. Thursday was Leadership Lab day, cadets wore uniforms all day, formed up, and drilled.

The Cadet Corps was organized like an Air Force Wing. The cadet Wing Commander and senior staff members were seniors. Juniors held Squadron Commander level positions. Sophomores and freshmen filled in the ranks. In ROTC, the cadets learned to follow, then to lead. The cadet Wing Commander and Staff held a weekly meeting and planned many corps activities.

The Corps met early mornings to accomplish physical fitness training.

Occasionally, they accomplished a "spirit mission." Once an entire ROTC hall floor was found covered with water filled cups. These young men and women were totally inspiring.

Cadet Spirit Mission

The program allowed the students to get a taste of the military while the system filtered the cadets. They had to succeed in college, pass background/drug checks, and pass the physical. That is why the numbers decreased through the school year as some are filtered out. And, of course, some just quit either ROTC or college.

One significant requirement was to compete for, be selected to attend, and then complete a Field Training camp. It occurred between the sophomore and junior year summer. The reason I taught the sophomore class was to know the cadets because I would have to rank them in the competition to be selected to attend a 4-week summer Field Training. There were four inputs that resulted in a cadet Order of Merit:

1. 50%, Unit number competing and My ranking
2. 20%, Cumulative Grade Point Average (GPA)
3. 15%, Physical Fitness Test
4. 15%, Test Equivalent Scores. SAT, ACT, or AFOQT-AA

Each individual cadets Order of Merit score competed against every other sophomore cadet from the 144 Detachments. Headquarters made a list from high to low score. They then rank ordered it and picked the number of cadets needed to fill the coming summers Field Training encampments. Those not selected were finished with ROTC.

I also used the Civil Air Patrol/ROTC program to get to know the cadets while showing them the magic of flight. It was a motivator for them, and a pleasure for me.

It was a great flight incentive program. As a CAP member, I flew most of the cadet flights. When I wasn't available one of the other CAP Squadron pilots flew them. I flew this program all three years I was in ROTC. It was great to get out of the office and fly! But more importantly, I was able to show the cadets the magic world of flight. It also enabled me to know each of them on a very personal level.

The cadet CAP members were issued flight suits. Renee made neck scarves with a US flag pattern. I wore one with my flight suit. When I met with the cadets for their first flight, I told them that they were out of uniform. Slightly shocked, they started to look over their uniform wondering what was wrong. I would pullout two of Renee's scarves and give one to each. Their big smiles were great as they put on the scarves.

Many of my cadets had never considered applying for pilot training before, especially the women. A number of them did as a result of the experience!

These flights were not pilot training flights, rather they were orientation flights. Usually I took two cadets, one in right seat and one in the back seat. I had two basic profiles. One profile flew to the west over the Colorado River and lakes. After leveling off I let the cadet fly the plane! They usually did pretty good. I would then do some mild maneuvers and air work. After an hour we would land at Burnet airport to change out the cadet's in the back seat to front. There was a small air museum at the airport which we toured. The new right seat cadet got to fly us back to the Mueller airport in Austin. The other profile was to the south landing at the San Marcos airport. There was a Commemorative Air Force unit with a museum there with a number of aircraft on display.

The joys of flying! One day I entered the visual landing pattern at the San Marcos airport. I heard the cadet in the rear seat gagging and getting sick. Now I couldn't just

CAP Incentive Flight

pull over to help her so I told the right seat cadet to dump all my stuff out of my green AF helmet bag onto the floor. He did and I handed my helmet bag back to the girl in the back seat and said, "Here, use this." And she did! We landed and taxied to the airport terminal and parked. She was a bit embarrassed but I told her no problem as it can happen to anyone. I did say that its tradition that whoever messes up the plane cleans the plane. The terminal folks gave her a bucket, rags, and cleaning materials. She did a great job cleaning the rear seat and rug. The three of us sat in the terminal for awhile then I told her it was her turn in the front seat and asked if she was ready to go. She flew us back to Austin just fine. I took my "used" helmet bag home. Renee kindly took it from me and cleaned it. Renee said, "The cadet had large shrimp for lunch." It was the first cleaning for the helmet bag in about 20 years. After that, Renee provided us medical vomit bags.

One Saturday I received the kind of phone call no Detachment Commander ever wants to get. The CAP plane landed and slid off the Mueller runway. The good news was nobody was injured. A CAP pilot was flying an orientation flight with two female cadets. I drove down to the airport to check on my cadets, the pilot, and the aircraft. I also needed to determine what had happened and write a report.

My report was this. The winds had changed to tailwind just as the pilot started the turn from base to final to land. It occurred so sudden that the tower hadn't had time to change the active runway to one that would have planes landing into a headwind. With the tailwind, the plane was travelling across the ground faster than normal. As a result, the pilot landed long and fast. The runway had a slight crest on it so it appears that you are closer to the runway end than you really were. The pilot pressed hard on the brakes. The right brake locked up and the plane was pulled off the runway to the right. There was a black rubber skid mark showing the locked right brake. The plane departed the runway to the right. It travelled across the grass for a short distance then hit a taxiway. The propeller hit the ground and the plane bounced into the air. It came down on the taxiway, crossed it, and then went back onto grass on the other side. The pilot didn't stop but continued in a slow taxi. The Tower Controller asked if he was ok. He answered with, "No assistance was required" and taxied the plane to the parking ramp.

In the end, the important thing is that nobody was injured. Since the propeller struck the ground the engine needed to be torn down and inspected. As for the two female cadets, one remained in the ROTC program, graduated, and went to AF flight school. She became a F16 pilot and flew combat sorties in Dessert Storm. She remains in the AF Reserve and is an instructor pilot.

My guess is that I may have shown her the magic of flight. Becoming a pilot was something she had never considered before.

As a CAP member, I flew a number of CAP border patrol missions. With another CAP member, we flew down to Laredo, or Del Rio, Texas. From Laredo we flew along the border to the gulf and back. From Del Rio we flew along the border to Presidio and back. While the pilot flew, the observer recorded any activity seen on the ground: location, number of people crossing the river border, vehicles, time, and day. The information was passed to the national border data base.

The Arnold Air Society (AAS) and Angle Flight-Silver Wings were two professional, honorary, service organizations that were affiliated with ROTC. The focus of the Arnold Air Society mission is to build strong officers for the United States Air Force. AAS members are cadets. Angel Flight-Silver Wings was in a military support Role made up of non-cadet college students.

Angel Flight-Silver Wings bid to become the National Headquarters was approved. I became the National Advisor. Later their charter changed and the organization moved out of its military support role into that of a collegiate advocacy group. While Arnold Air Society and Angel Flight-Silver Wings became separate organizations, they maintained a strong partnership that is based on highly complementary missions. Years later the name was changed to "Silver Wings."

AAS/Silver Wings held an annual National Conclave. This year it was held in New Orleans. I flew the CAP Cessna 172 with 3 AAS cadets into Navy New Orleans, Naval Air Station. We landed and checked into billeting. The four of us meet at the desk and told the lady we would walk to the river and ride the ferry over to New Orleans. I thought her reaction was strange. She said with a shocked voice, "You're walking?" I assumed she thought the distance might be too far.

The four of us were about half way to the river when we came across a group of 12-15 young men walking the other direction. They roughly pushed through the four of us. I was hearing comments like, "You must be undercover cops or you wouldn't be here." We continued to walk. The plan was to spend the first night at the Naval Station then move into New Orleans hotel the next day. I now realized why the lady at the desk showed shock. The issue wasn't distance, it was safety. I told the cadets under no circumstance were they to make the walk again. Use a taxi.

The convention was a success and fun. We flew back to Austin.

Field Training, Lackland 1

One of the ROTC cadet filtering events on the path to an Air Force commission was being selected to attend a four-week Field Training event. The camps occurred in the summer after the sophomore year. There were numerous camps at military bases spread across the country. I was assigned as Commander of Field Training camp "Lackland 1."

This would be a test camp, the largest camp ever. Years ago, the one I attended contained 250 cadets. There would be 520 in the camp I commanded. I travelled to Lackland AFB in San Antonio, Texas, a week before the cadets would arrive. Twenty-two officers plus enlisted converged from ROTC Detachments across the country. I had a week to get organized and build the staff into a team.

520 AFROTC Cadets in 4 Week Camp

The ROTC camp I attended years earlier was all male. In this camp, there were females in each flight. The females didn't room with male flights, it wasn't coed.

The four weeks was a true challenge. The cadets had a bed time and get up time, we didn't. The first morning the Flight Surgeon came in and told me that a cadet had chicken pocks. He wanted to quarantine the entire camp! I pushed back. She arrived late, after all the other cadets, so her exposing them would be minimum. Finally, the Flight

Surgeon agreed but blood tested the few cadets she was in contact with. I had to send two cadets' home. We put her into the Lackland base hospital. She recovered, then stayed and joined the next camp, Lackland 2.

It was amazing to watch the cadet performance improve through the month. The month could be described as a chaotic first week, a much-improved second, rolling along on automatic third week, and the final week, I'm wondering why we are still here.

The first morning assembly where the camp fell into formation was total chaos. Flight leadership positions changed frequently giving a leadership opportunity for many cadets. They marched to every activity. There was a monster obstacle course that they completed a few times. Calisthenics occurred daily where cadet's results scored. The flights were unbelievably competitive. We went to the firing range and shot the Beretta.

I was amazed at how many cadet grandparents died during the month. I would get a Red Cross communication about a death. The Chaplain joined me to meet with the cadet. He asked me to tell the cadet about the death and he would console them. I would call the cadet into my office and gently inform them of the death. I offered a phone call to their parents and said they had an option to leave for three days to attend the funeral. They all called but none left.

Flight waiting to enter the Chow Hall

The camp occurred in May-June time frame and it was Texas hot! The cadets were continually active. Even though we pushed drinking water on the cadets many would get dehydrated. We had a number pass out resulting in many visits by the base ambulance. Whenever the temperature and humidity combination exceeded a critical number, we stood down. I always felt relieved whenever a responding ambulance passed by the camp for another destination.

Completing the camp was a requirement to continue in ROTC. I had one young man injure a leg and needed to use crutches. In theory, I should have sent him home but I didn't. I asked him to follow his flight along as best as he could. About a week later he was off the crutches and could keep up with his marching flight. He completed the camp.

Years later, an officer approached me asking, "Do you remember me?" "No, I don't, but should I?" "He said, "You were the ROTC camp commander at Lackland 1. I was the cadet using the crutches." I remembered him! He then thanked me for letting him stay to complete the camp.

The month-long training schedule was intense. Lackland 1 ended and we all went home. I was totally exhausted. It took me a full week to recover. I'd sleep all night, then get up have breakfast, then fall asleep again.

First AFROTC Dining Out

Every year the cadets held a Dining Out. That is a formal military dinner that includes nonmilitary folks, wives, and friends. A Dining In only include military members. At each dinner there were Air Force traditions. There were Rules of the Mess. Violate a rule of the Mess and you could be sent to the Grog Bowl. There were also Rules of the Grog Bowl. The Grog Bowl could contain a fowl concoction. (See Appendix 4)

As we ate dinner, cadets would stand up and accuse another cadet of violating the Rules of the Mess and explain why. If the President of the Mess agreed, then the violator would be sent for a drink from the Grog Bowl. It was great fun.

This was my first Dining Out at the ROTC detachment. I invited a University Professor to be the speaker. He invited a couple of friends from England. Did I mention that my guest speaker was once the Secretary of the Air Force and once Chancellor of the University of Texas System?

A cadet stood up and pointed out another cadets' infraction of the Mess Rules. The President of the Mess sent the offenders to the Grog Bowl. It was uncovered for the first time. It was a toilet bowl. I was both shocked and totally embarrassed. My guests were watching cadet's drink from a toilet!

About two weeks later a message from ROTC headquarters came down telling all ROTC detachments to not use toilets as grog bowls. I'm sure my guest speaker caused the message.

In September, 1996, Renee and I attended a 4-day Air Force Association (AFA) National Conference/Arnold Air Society (AAS) Executive Board Conference in Washington DC. AAS cadets from across the country attended.

One evening, Renee and I, along with the Detachment Commander from in San Antonio (Patty), accompanied about 15 cadets downtown into DC for dinner. We rode the Metro train, then walked a few blocks to the Hard Rock Café. We had a rousing time and a great dinner. It was getting late and we needed to leave to catch the last Metro ride to return to our hotel.

We left the Hard Rock, walked a block and turned left at the corner. At the next corner there were two drug dealings going on. Two guys were heads down into the two cars. As we approached, their heads came up and locked onto us.

Renee, Patty, and I sensed danger and started to encourage the cadets to move quicker toward the entrance to the Metro. I could see the lit Metro sign just one block ahead. We arrived at the Metro escalator and started packing cadets onto it when Patty cautions, "A man is coming toward us." I'm last, the others disappear down the escalator.

A black man arrives. He and I are face to face, within an arm's reach from each other. This encounter was brief. He threatened me, "Give me your wallet, mother f…er, or I will shoot you." The man just threatened to kill me. We are eye to eye. I note that his eyes are clear, so I didn't expect any irrational drug caused behavior. Using peripheral vision, I see he is wearing a military green fatigue jacket. His hands are in the pockets. I have a fatigue jacket so know what could be in the pockets. He was as tall as I was, younger than me, and stocky.

I quickly ran through my options. A physical fight with him wouldn't be a good option. I couldn't wrestle with him. A palm of hand strike to his nose would push his nose bones into his brain. A fist to his throat could crush his air way. I hadn't seen a weapon yet, so if I missed, I would have to fight him. I decided to fight only if he became physical with me.

I had not seen a gun so I said, point blank "I'm out of here," I turned around, took one step forward onto the escalator, and started riding it down. I didn't look back. The first few feet I'm thinking, I'm going to get shot in the back and tumble down the escalator into the mass of cadets below. About 20 feet down I figured he wouldn't get anything by shooting me since my body would end up at the bottom of the escalator. Finally, he probably didn't have a gun.

Renee and others were at the bottom of the long escalator looking up. Renee said it was like a slow-motion movie waiting to see my face appear as the escalator slowly came down. No Rick. Then my feet, legs, and me. I still have that wallet.

Unexpected challenges occurred frequently during my life, I was sitting at my ROTC office desk and get a phone call from the Air Force Casualty Center located in San Antonio, Texas. The voice said, "I'm sending you a fax about an AF casualty. A female Air Force Reservist died in a C130 aircraft crash in South America. You have three hours to notify the family."

Out of respect, I drove home and changed into my more formal Class A uniform. I've lived with, faced, and known death in my years in the military. Friends have died. This was different. Have you ever walked up to a stranger's house, knocked on the door, and told a family that their daughter died? I have. How sad.

The big event for graduating seniors was getting their assignments. I would announce what they received. We made oversized cardboard pilots and navigator wings to pin on their chests. I had a friend assist awarding their assignments. (John had survived 5 years as a POW in Vietnam.) I announced the name, the assignment, and he pinned on the wings.

I totally enjoyed my ROTC time. I proudly sent many outstanding cadets through the official selection process into the air force. Having said that, I selected one outside the system. Here's the story:

Know what you want and act. Congress sets the yearly end strength for the Air Force. That is, how many enlisted and officers could be on active duty at the end of each year. That in turn drives how many cadets are offered an Officer Commissioning allocation.

A commission is not a guaranteed thing. It is actually an extremely competitive event. Each cadet effectively competes with every other cadet within the 144 colleges for an allocation.

Field Training is one event that all cadets had to be selected to attend and successfully complete. Camps occurred during the summer months between the sophomore and junior year.

The ROTC Field Training selection process is competitive within the Detachment and across the country, each cadet was measured in four areas, and the math formula resulted in a final score, that final score was ranked against all the other cadets across the country. HQ ROTC set the number of cadets to attend FT based on model that provided for attrition. Those cadets scoring above the final score were assigned to attend a Field Training Camp. Those below weren't.

The results came back and Dave was not selected to attend Field Training. About 30 minutes before I would announce to all the assembled cadets who was selected, I took him aside and explained to him that he wasn't selected. I didn't want him to be surprised. The detachment actually had a very high selection rate.

He was an outstanding young man. Later we talked. I offered to speak with the Navy and Army ROTC commanders to see if he could join them. Dave said, thanks but no, he only wanted to be a pilot in the Air Force. His dream over, or so he thought.

A couple months later the school year ended, it was summer, and our cadets began to depart for various field training camps. I get a call from ROTC Headquarters mid-morning. A cadet somewhere had dropped out and one cadet was needed to fill a camp. Did I have one available? If I did, the cadet needed to travel to camp that same day! I asked for 30 minutes to answer and hung up. I immediately thought of Dave. I asked my staff what time the cadet needed to depart Austin on an aircraft, 5 pm was the answer.

I called Dave on his home number. Luck was with us and he answered from the kitchen. I explained the situation to him and then asked him if he wanted this? I told him he needed to be in the Detachment by 3 pm for a 5 pm flight. Without a moment of hesitation, he thinks out loud: "I'll give my accounts to Mary (he sold real estate) and I'll leave a note on the table for mom telling her I'll be back in a month. Yes, I'll be there."

He took no time to decide to go, it was instantaneous. At this point his Air Force dream was gone. He had no expectations to attend. After a 2-minute call from me he was on his way. He knew what he wanted, took the opportunity and ran with it. A man who knew what he wanted!

I called ROTC HQ telling them I had this covered. About 2 pm Dave came into the detachment with two half-filled suitcases and four Target shopping bags. He put it all on the floor and packed his suitcases right there on the floor. My NCO drove him to the Austin airport and off to camp he went. I called ROTC with his info and travel plan.

Dave competed very well at Field Training. He eventually was commissioned as a 2nd Lieutenant. Last I heard, the young man who wasn't selected to attend summer Field Training is still a fighter pilot in the Air Force! He is living his dream! I selected him! Not the Air Force ROTC system.

My three years over, I was 28 years in the Air Force. I had 2 more years before reaching the mandatory retirement at 30 years, I was assigned back to the Air War College Faculty.

Chapter 12 - Air War College Faculty, again

I returned to the Air War College faculty. We were given a base house just three doors down from where we lived during our first tour there. The floor plan was the same except it was reversed. All our furniture went back to the same locations that they were during the first assignment.

Since I had been there before, I already knew the ropes. I was assigned as Director of the Regional Studies Program. As such, I oversaw the coordination for 230 travelers on 14 separate trips to 45 countries.

I also was the Seminar 19 Director. Each Seminar contained 13 students of all branches, including two foreign officers.

Most US military schools have international students attending. They add immensely to the event. They learn about us while we learn about them. These relationships have proved valuable through the years. Friendships are made. When I was a Student at the National War College there was an officer from India in my committee.

There were two international students in seminar 19. One was from Taiwan, Col K. He told us his nickname was "Indian." The other student was from an Arab country. During the 10 months we came to know each other very well. They were both outstanding, awesome men.

Renee and I lived on Maxwell AFB and invited the seminar to our house for a number of seminar parties. One was a Halloween party. It was great fun. This was a new experience for the Arab family. We all had dressed up in costumes, went trick or treating, then bobbed for apples. Ed was a U2 pilot. He dressed up in an elaborate witch's costume that he made. He sat on a rocking chair next to our front door. Kids would come up to look at him, then touch him, he moved and moaned. The kids would scream and scatter in terror! Scared them big time.

Toward the end of the course, the Arab man invited about 8 of us to his apartment for dinner. The dinner was superb. He didn't sit and eat with us, rather he just served us dinner. This was his way to honor us.

Sometime later in the course, I was alone in the seminar room with him. I was leaning over reading the next course period material. He said, "Rick, I wish I had not brought my family here." I knew what was next but asked, "Now why would you say that?" He said, "My family doesn't want to go home." He had two daughters and a son, all teenagers. Here was a Colonel in his countries military and his family rejected that country.

The Taiwanese officer invited us to a dinner at an Asian restaurant in Montgomery. He arranged for a special dinner be prepared that wasn't on the restaurant menu. The food just kept coming, course after course. It was terrific. He also brought a number of bottles of Taiwan Liquor. As we ate, "Indian," said each male needed to make three toasts. Toast made, then a shot of Liquor was knocked back. There were 4 males so 12 shots! During the toasts we asked a number of times if the ladies could participate. Each time he firmly said, with a hand wave, "No!" That was a good thing. He gave each of us a gift bottle of the booze. I still have the unopened bottle of the Taiwan Liquor. (We named it, JP-4, which is jet fuel.)

After dinner things became fuzzy and I'm less able to recall much. Having said that, we went to a club, somewhere. I recall there were tables and chairs, a dance floor, and a stage with a female singer and band. At some point, I looked over at Ed, the U2 pilot. He didn't smoke, but there were lit cigarettes sticking out from each ear, two out of his nose, and about four from his mouth. I recall looking up at the stage to see "Indian" on it singing and dancing with the singer. Our ladies drove us home. It was a quick night. We were home by about 10 pm. The next day there wasn't really a hangover per say. I described it as an "Out of body experience."

When I was at Maxwell the first time, I completed all the academic course work for a Doctor of Public Administration (DPA) degree from the University of Alabama. All I needed to complete was to write and defend a dissertation.

While at the University of Texas for three years, I submitted several dissertation proposals to my Alabama professor. He rejected them for various reasons. I became involved in my ROTC assignment and stopped submitting. Now back at Maxwell, I attended what was an Alabama DPA cheer leading meeting. There were about 30 attending. As I sat there, I decided that I wouldn't continue. I returned to the house, sat Renee down, and told her I was done with the degree.

During these last two years at the Air War College, I travelled on two Regional Studies trips: The first was the NATO trip to Europe. We travelled through Paris, France; Brussels, Belgium; Stuttgart, Garmish & Munich, Germany.

Normandy Beach

I walked on the Omaha Beach where my uncle landed on D-day during WWII, saw General Patton's grave, ate mussels in Brussels, and toured Dachau a WWII Nazi extermination camp. My Dad fought across Europe.

Ovens in Dachau

My second trip was to South East Asia. We stopped in Hawaii for briefings, then travelled to Bangkok, Chiang Mia, and Chiang Ria, Thailand; Hanoi, Na Trang and Ho Chi Ming City, Vietnam; then finally to Singapore.

I had been to Thailand flying combat sorties from Utapao during the SEA war. My hair stood up when our KC135 entered Vietnam airspace, the last time there was during the war. After we unloaded, the tanker returned to Thailand. We arrived in Hanoi at night. We were being hosted by the military. Dinner was spectacular. I felt guilty

Hanoi Hilton

eating what was a banquet. I recalled how poorly US POWs were feed. After dinner, two of us walked a few blocks to the infamous Hanoi Hilton prison where US prisoners were kept and tortured. Tortured to tell the world how good they were treated. Think about that for a moment. There were very few people on the streets at night. The two of us took turns standing against the outer wall taking pictures of each other when a door swung open, a man leaned out, and motioned for us to come in. The two of us looked at each other, then went in. The man puts his hand out rubbing his fingers together meaning, give me money. We paid him some Dong (Vietnamese Currency). They had made the prison into a museum showing the time when the French controlled the country and Vietnamese were prisoners. I slowly walked past the prison cells. I knew pilots that survived in these cells for years during the Vietnam War. It was a somber reflective walk.

Prison Cell Doors in the Hanoi Hilton

I observed significant differences between North Vietnam (Hanoi) and what was once South Vietnam (Saigon). In the north folks wore the typical cone shaped head gear and rode bicycles and some motor scooters. The streets contained a continuous flow of bicycles. The embassy folks briefed us on how to cross a street. Once you step off the curb, keep walking at the same speed. Don't stop, change speed, or turn around. The bicycle traffic would flow around you.

Hanoi Traffic

We visited a Museum at the Hanoi airport. There was a display with a B52 being attacked and shot down by fighters. I told my escort it didn't happen, surface to air missiles (SAM) yes, fighters no. He just smiled. There were still bomb craters off both ends of the runway. We rode the Vietnam Airlines to Na Trang, then to Hoi Chi Min City (Saigon). The Na Trang beaches were beautiful. In Hoi Chi Min City, the street traffic was filled with cars. Folks wore more western cloths. We were told not to ride any of rickshaws because they would probably take us off on a side street and rob us.

I was walking along the street. A young girl appeared and wanted me to buy post cards from her. I said, no. She followed me as I walked from store to store. As we walked along I asked, "Why aren't you in school?" She said in excellent English, "I'm out of school for the rest of the day." Then I ask," Does your mom know what you're doing." She points across the street. Her mom was on the other side of the street walking along with us. I bought some post cards from her. The countryside was beautiful, the people friendly.

The irony of this trip was that I served 30 years in the military. My first years were fighting in SEA, my last year I was visiting SEA.

Civil Air Patrol

While at Maxwell, I joined the base Aero Club again and flew a variety of single and multi-engine aircraft both privately and as an Instructor Pilot. I earned the Certified Flight Instructor Instrument (CFII) rating so I could teach instrument flight.

I also joined the Civil Air Patrol Squadron located on the Base. The CAP Cessna 172 was located at the Montgomery airport. The reason it wasn't at Maxwell was that Maxwell shutdown runway operations at night and the CAP plane needed to be available 24/7. I became a CAP Instructor and Check pilot.

My initial CAP check flight was to fly a 10-line search pattern north of Bessemer, AL. On aircraft startup the aircraft GPS avionics didn't work! There were two CAP members in the plane. The check pilot sat next to me and the local Squadron Commander in the back seat. I took off and flew north using a hand-held paper aviation map. I navigated to the search area then visually flew three search lines out of the planned 10-line search area. The check pilot announced, "That's enough." Both of my CAP ride along evaluators were totally stunned. They asked me, "How did you do that? We couldn't have found the search area without aircraft GPS." I said, "I've flown many hours at low level using visual maps." I knew I passed.

At the Maxwell Squadron I flew pilot proficiency flights, gave check rides, and flew Emergency Locator Transmitter (ELT) training sorties. The CAP is a search and rescue operation. Back then, most civilian aircraft had an ELT attached inside the tail. When a plane crashed the impact G-load would activate the transmitter which gave out a continuous radio signal. When heard, perhaps by another airborne aircraft, Air Traffic Control would be contacted to report it. ATC would contact the National Rescue Coordination Center. They, in turn, would contact the local CAP Squadron to search. The ELT transmitted continually until the battery ran down.

CAP crews are trained to home in on the ELT signal to find the crashed plane by flying a number of lines noting where the signal was strongest. The CAP plane was crewed by a pilot and two observers. One observer was in the front right seat scanning to the right; the other in the left rear seat scanning the left side. They searched hoping to see the crash location.

We trained by having a member take and place an ELT simulator out away from the airport 10-15 miles. We then flew practice sorties homing in and locating it.

Technology has changed the scenario. New aircraft equipment now transits the GPS coordinates via satellite directly to the National Rescue Coordination Center. Exact location now known, rescue efforts respond.

I flew as a CAP member for a year when the Squadron Commander approached me and asked if I would take over for him as Squadron Commander. He was beginning work on a law degree and didn't think he could do both. I said yes. I was CAP Squadron Commander for my remaining one year at Maxwell.

This was my first experience as head of an all-volunteer organization. I had run into many folks around the area that knew about CAP but didn't join. I'd ask why not? Usually they said that they had attended a meeting but nobody paid any attention to them so they just didn't go back. I had a fix for that. Whenever I had a Squadron meeting and I noted anyone new attending I would acknowledge them during the meeting, answer any of their questions, then assign our personnel rep to give them an application to fill out right there!

We would spend time with the person after the formal meeting and handed them the orientation materials needed to complete their membership. The next day I hand carried the new application over to the CAP Headquarters also located on Maxwell. I noted over time that more and more new folks attended meeting and joined the Squadron. During that year I grew the Maxwell Squadron to be the largest CAP squadron in Alabama.

A funny story. My squadron met in a building that was located among a number of buildings on the north side of Maxwell AFB. There was a road running between the buildings next to us that led to a Federal Prison. During a squadron meeting I was conducting training in the use of the ground ELT locating equipment. It is a hand-held antenna/receiver that is swept round searching for the strongest ELT signal. The intent is to home in on the aircraft crash site. As I went over the theory of operation, I had a CAP member take a training ELT transmitter outside to hide so after my class we could go outside and use the handheld receiver to find it. The class ended and we went outside.

As we exited the CAP Squadron building there were about 10 security police cars and an ambulance arrayed in a distant sweeping arc in front of our building. A security officer motioned to us vigorously to come to him. We did. I asked him, "What's going on?" He said, "A prison security officer travelling to the Federal Prison saw what might be a bomb." Well, it turns out that the maybe bomb was our practice ELT. My CAP member had put it down next to a building and had put a cardboard box down on top of it with the transmitting antenna sticking out the top. The prison security officer saw it as suspicious and called in a possible bomb threat.

I told the officer what it was and volunteered to retrieve it. He said ok. I noted that he didn't go with me to get it. I took the cardboard off the training devise and showed it to security. They were happy and departed. I guess the CAP Squadron made it onto the security police blotter that night.

My last two years in the Air Force passed quickly. The Air War College held a wonderful retirement ceremony for me on 6 August, 1999. My dad, sister (Bev) and her husband (Tim) attended. A special treat, two of my 1969 ROTC classmates (Dave and Scott) also attended. The huge room was packed.

My 30-year history was reviewed, a retirement medal awarded, and Renee recognized. Finally, it was my turn to talk. I recognized and thanked everyone for their support and cooperation through the years.

Then I recognized my Dad and explained to the attendees that during WWII, he fought across Europe in a half track armed with quad .50 caliber anti-aircraft machine guns. He was attached to General Patton during the Battle of the Bulge. I asked Dad to stand, he did. The room erupted in applause. That was the high point of my retirement ceremony.

I retired on 30 August 1999, after 30 year and 7 days in the Air Force.

Chapter 13 – Photo mapping job

Retired, we moved back to our house in Round Rock, Texas. For three months in my life I was unemployed. This was a new experience for me as I have worked from high school until my retirement from the Air Force. I was learning how the civilian world worked as I applied for a number of positions around the Austin area.

In the meantime, I painted the exterior our house, then crossed painting houses off my list of things to do.

Reading the Austin newspaper, I came across an ad for a pilot to fly a photo mapping aircraft for Atlantic Technologies, headquartered in Huntsville, AL. The plane (a

Twin Cessna 340) was located at the Austin airport. That ad was strange because most flying jobs were advertised in other flying related publications and not the local newspaper.

Curious about the ad, I called the contact number and talked with Jim. I said to him his posting in a local paper was unusual but I told him I could fly a straight line. We talked, then met. We flew together and I demonstrated to him that I could fly the needed precision; like a B52 bomb run. He hired me. (I had been an Instructor Pilot at Maxwell AFB where I flew a twin Cessna 310, slightly smaller than the Cessna 340.)

A week later I arrived at the Atlantic Technologies office in Austin for my first day. Jim had left the company to fly an eye surgeon around Texas in a Cessna 340.

Through the years, Jim and I have remained great friends. He flew for the eye surgeons for awhile, then moved to a charter company. The charter went out of business while he was out on a charter trip. They closed down operations and didn't even inform him! He was stuck on the road to find his own way home. Jim was then without a job! Such is the commercial flight world.

I went from a B52 Bomber Pilot (long range demolition) to a reconnaissance pilot (taking pictures).

Twin Engine Cessna 340

After 30 years in the military this was my first civilian job. I learned the photomapping operation on my own.

Have you ever seen an X mark painted on a highway? That X is a precise GPS measured location. There are two more locations X marked about 2-300 feet to the right and left off the road that you don't see.

I flew anywhere from 700 to 17,000 feet above ground depending on the job. The photos were taken with a 30% overlap of pictures. The big camera hung out the bottom of the plane. Our 12-inch-wide film was developed and then computerized. The project was then linked to the precise GPS coordinates (the X on the road).

The time interval between when the camera would fire depended on the flight altitude. At the lowest altitude it could fire every 7 seconds while at high altitude it might be every 25 seconds. I was hand flying and needed to be wings level when the camera fired. There couldn't be overcast or any cloud shadows in the photos. The sun angle was also important so all flights were conducted between 10 am and 2 pm.

I remember explaining this new flying job to Renee's dad. I told him it was a great job because I only had to work between 10 am and 2 pm. He gave me a big belly laugh.

Neil Running the Camera

Neil was my project planner and airborne camera operator. He was from Warton, Texas. We once landed there and I met his family. Neil planned each job and loaded it into the onboard GPS computer that gave me inflight guidance over the project. Neil was a perfectionist wanting everything on each job to fly perfect. He was the kind of man you wanted to crew with, one you trusted.

We had a long list of projects planned that were located all around Texas. I watched the weather to determine where the weather was suitable to fly and off we would go. There were two photomapping planes based in Huntsville, AL. Neil and I produced more film and projects than the two in Huntsville did. I think it was because we were autonomous. We controlled our schedule. The Huntsville crew had to deal with the staff there.

While most jobs were in and around Texas we also had projects to fly around the country. Most were for road projects for the Texas Department of Transportation. They could use our product to determine how much earth needed to be moved in or out to work on road projects.

We also flew projects in Alabama, up to Chicago, and to New Orleans. We flew over rock quarries on an annual basis. The quarry company could use them to determine the volume (amount) of material removed since the last overflight. We usually could complete a rock quarry flight with just three quick overlapping photos. We flew over road projects, entire counties, limestone quarries, and coal piles in Texas City.

Pilot Guidance System

One project was 8 counties around Houston. It was huge. We flew it at 17,000 feet. Unpressurized, we wore oxygen masks. Legs were north/south and 40 minutes long!

The camera fired every 25 seconds. It took months to complete it. Then Huntsville told us that we needed to retake 3 photos. We needed to fly to Huntsville for a meeting. I planned to over fly and retake the three Houston photos, then fly to Huntsville. It was tight fuel wise but doable.

After we shot the three photos, I requested a direct flight to Huntsville from Houston Center. I was given a heading to fly and it wasn't toward Huntsville. I requested direct to Huntsville a number of times. I watched our fuel situation deteriorate. Finally, we were cleared direct. I calculated that we would arrive with just enough fuel to meet FAA requirements.

Inflight winds changed and I was wrong. We arrived with just enough fuel for one approach and landing. We flew into the Huntsville area and listen to the landing

At 17,000 Feet and Breathing Oxygen

weather broadcast. There were low clouds with runway crosswinds that exceeded the demonstrated maximum of the Cessna 340.

I didn't have fuel to divert, I had to land on the first attempt, so I flew the instrument approach to 200-foot minimums. Seeing the runway, I fought the plane in the crosswinds onto the runway. The landing wasn't pretty, it was really rough, but we survived to fly another day.

Two of most challenging projects, for different reasons, were flights over the City of Irving, Texas, and a three-shot rock quarry in Cancun, Mexico.

What made the City of Irving project challenging was that the area extended from over the Dallas/Fort Worth (DFW) airport to the east toward the Dallas Love airport (DAL). The job planned out would be 12 north/south lines at 4,000 feet. Each line was 10 miles long. The problem was that I just described the airline arrival/departure routing at both DFW and Dallas Love airports. To accomplish this project, we would be among the airline traffic! This probably wasn't something that the DFW air traffic controllers were going to agree to.

I called the Dallas approach control phone number and explained what I needed to do. He just laughed at me and said there was no way that we could ever mesh in with all the airline traffic. I asked him to sit down and to just humor me for a minute. I had faxed him a map depicting the 12 lines I needed to fly. I asked him, when was the slowest air traffic load time frame at DFW? He said that 1145 was the end of the morning inbound airline rush. I asked, what if I was airborne to the south of DFW out of the Redbird airport in a 4,000-foot orbit. I would check in with them and if the traffic load permitted, they could clear me for some photo runs. He said ok!

On the planned fly day before 1145 I was airborne south of DFW over the Redbird airport in an orbit at 4000 feet. As I listened on the approach control frequency, I could hear the air traffic chatter decrease to nothing! I called approach control and asked for clearance. He cleared me to fly line 1. Line 1 flew north 10 miles directly over the DFW east passenger terminal. I was then cleared to the east for line 4 flying to the south. Then I was cleared for line 2 to the north. Line 2 was still over the DFW airport runways. Finally, as we flew line 3 to the south the controller said we were done at the end of the line 3. I asked if we could fly any lines farther to the east, a line closer to Dallas Love airport. He said, no I was done. I thanked them for their help and landed at the Redbird airport. Four lines flown, eight to go!

The next day the weather was clear and I flew the remaining 8 lines! Job complete! What were the chances? The following days storms covered the area. I was very lucky to get this job flown!

The second challenged flight was for an annual flight to Cancun, Mexico. This was a different challenge for a different reason. We needed to fly short three photo shot line over a rock quarry south of Cancun, Mexico. Neil and I flew into New Orleans. We rented a survival kit and life raft from the Fixed Base Operator to be legal to fly over water. We spent the night in New Orleans. And yes, there is beer and awesome food to be found on Bourbon Street. The next day we flew over the Gulf of Mexico and landed in Cancun.

To fly into Mexico, you need liability insurance that is specific to Mexico. Upon arrival at the Cancun airport, a Mexican Inspector reviewed our paperwork plane side. Our aircraft insurance paperwork was wrong! The paper work said the tail number was 27642, but the aircraft actual number was 2764Z. The last letter was a 2, it should have been a Z. He looked at it and said that it was ok with him, but it might be a problem with some of the other Mexican Inspectors. He left.

We went into the terminal and immediately called the insurance company in the US. I said I needed the tail number corrected and the agent immediately faxed to me a copy of the corrected insurance paperwork. I put the corrected insurance paper work into the plane.

Bourbon Street

Neil and I flew a number of days trying to make the short three shot photo run, but each day the weather wasn't right. There were always clouds covering the quarry. This was painful because each day we planned to fly the three-shot job and fly back to New Orleans. That meant each day we checked out of the hotel and turned in the rental car.

Finally, we took the three quarry pictures needed and landed back at Cancun! I filed the flight plan to return to the US. We were in the aircraft to preflight the plane, there was a knocking on the side of the aircraft near the back door. It was a Mexican Inspector. I go back and exit the plane. He asked for our paperwork. I give it to him knowing he was looking for the incorrect paperwork. I watched him carefully as he looked over the insurance papers. He looked at it and his eyebrows kind of scrunched together. I can see he doesn't see what he expected to see and that the paperwork is correct. He looked at the paperwork for a minute thinking, then he looked up and he accused us of violating Mexican flight rules the previous days and we were to follow him.

He led us into his Cancun airport office where he sat down behind the desk. Neil and I are standing in front of it. He started telling us about all the flying offenses we committed. I know this is all BS. (I revert to my Air Force Prisoner of War training experience and don't believe anything I see or hear). He lectured and threatened us for about 20 minutes. He then left the room briefly and Neil says "Come on let's just go." I told him, "Stay put and don't say anything, this is their show, and I don't want to end up in a dirt floor jail cell."

Shortly, two different guys show up and told us that our fines could be as much as $80,000. Then the original Inspector returned. One of the two men said they wanted us to go with them to another location.

Now, I've been around the world a bit. This, in my mind, could be a life or death event that called for action. When people want to take you somewhere the situation isn't good. I don't even know if they were actual security personnel or whoever. If we left with them, nobody, and I mean nobody, in the states would know where we were, who had us, or what had become of us. It was time to act.

I said, "Just a minute." I pulled out my telephone calling card out of my pocket, turned, and quickly walked out of the room to the phone hanging on the wall just outside of the room. I half expected to get tackled by the men. I didn't even look back to see what their reaction was. I dropped the card into the phone and dialed the company office in Huntsville, AL. A lady receptionist answered. I said, "My name is Khalar, I'm in Cancun. Don't put me on hold. I need to talk to the CEO, now."

He came on line with a, "What's going on?" I tell him we are in Cancun airport, the situation, and that some men want us to go with them somewhere. He says he will start making some phone calls and to call him back if we are released.

I hang up and turned around to see the Inspector standing just behind me. The other two guys were gone.

They heard the discussion. They knew they just lost the pressure edge of the threat that we could disappear since now my people now knew where we were and what was going on.

His demeanor now totally changed. The Inspector now sheepishly said, “For $200 I can make your problems go away.” I said, “So for $200 we can leave?” He replied, “Yes.” We went back into his office. I start pulling $20 bills out from my wallet, socks, pockets, and underwear. As I leaned across the desk handing him the $200 I ask him a question that I already knew what his answer would be. I ask, “Can I have a receipt?” He waves his hands in front of him in a slight panic saying, “No, no, no!”

Neil and I go out to the plane, start the engines, and taxi out. Just before we get to the departure end of the runway the ground controller told us to stop and hold our position. I didn’t see an operational reason for this request as there was no activity anywhere near us. I suspected the stop might be connected in some way to the two guys, or perhaps a discussion was ongoing about who gets a portion of the bribe money.

As we sat there, I told Neil, “Neil, if we are asked to return to the parking ramp, or if a car drives out toward us, I will takeoff without a tower clearance and fly back to the states.” We sat there for about 5 minutes then were cleared for takeoff.

I often wondered if there was a bribe sharing agreement between the Inspector, the two men, and perhaps even the tower personnel. I missed an opportunity to screw with him. I wish that as we flew out of the area I had announced on the tower frequency that I hoped the Inspector enjoyed the $2,000 bribe we gave him. If they had a bribe sharing agreement then maybe he would have had to explain to them where the other $1,800 went. A missed opportunity.

BTW, this occurred in 2000, I’ve never been back to Mexico. I’ve told this story to dozens of pilots. I wonder how much revenue loss this criminal Inspector with his threats, intimidation, and $200 bribe caused Mexico to lose.

Chapter 14 - Chief Pilot, University of Texas System

One afternoon, I was in the Cessna 340 photomapping aircraft getting ready for a photo flight. A friend was walking down the ramp, stops, and we talk awhile. He mentions that the University of Texas System is looking to hire a King Air 200 pilot.

I called the System and asked about how to apply. I was told I needed to stop at the downtown office complex and pick up a paper application. I did, I filled it out with my sturdy pen and submitted it along with a resume. That night I called Bud, the retiring 70-year-old pilot. It turns out he was also a retired Air Force pilot. We discussed the UT System operation, flight frequency, pay & benefit issues, and the aircraft. Bud was extremely helpful and supportive.

Preparing for interviews, I studied the UT System. I researched its structure, the organizational chart, the staff members, and who the Regents were. I learned that the UT System is huge! It consists of 8 Universities and 6 Health Institutions spread across the state. The UT System staff numbered around 400. They worked all sorts of issues.

The plane was used strictly for official business. The reason for the flight request and passengers list were carefully screened before approval was given. Also, since the System was collocated with the University of Texas at Austin, occasionally the University President, staff, and athletic personnel used the plane.

I accumulated an extensive information folder. After my first interview with the man that would be my immediate boss, I showed him my research folder. He was totally impressed with my research.

I was scheduled to interview with the Chancellor. There were about 10 Vice Chancellors, my soon to be boss, and the retiring pilot in the room. The Chancellor and I sat on a couch, he asked me to tell him about myself. I talked about my Air Force and flying history. After awhile he asked, "How long have you been flying the King Air 200?" I answered, "I've never flown one." He was visually taken aback a bit, then asked, "Why would we hire someone who has never flown one?"

I sensed that this could be the make or break answer. I revert to an instructor explaining, "Well, all aircraft fly in accordance with the same flight rules. The principles of flight are the same. Aircraft controls, flaps, ailerons, rudder and elevator are on all planes. All aircraft have various types of fluids such as fuel, oil, hydraulics that are stored and utilized. What I need is training about the aircraft cockpit configuration." Then using my hands as if they were on the flight controls, I said, “Finally, push forward, trees get big, pull back, trees get small.” With that comment, the entire room burst into laughter. I knew that I had survived the question.

When I applied, I provided several references. A couple years earlier, for 3 years, I commanded the AFROTC Detachment 825 located on the University of Texas at Austin. I met a professor there. From time-to-time, he would come to the detachment and lecture the cadets. He had once been the Secretary of the Air Force and also had once been the UT System Chancellor! My soon to be boss later told me he called my reference for a recommendation. He happened to be at the Pentagon. My boss asked him for a recommendation. The response was simply, “Hire him.” It’s good to have friends.

I was hired. Later, I learned that 11 pilots applied for the position, 5 were considered, 3 recommended, and 1 (me) interviewed with the Chancellor. Of the 11, I was the only pilot without any KA200 experience! Some that weren’t selected were King Air 200 pilots flying out of the Texas State Aircraft Pooling Board. Later it became Texas Department of Transportation Aviation Division.

I flew a few flights with the retiring pilot (Bud) then attended SIMCOM King Air 200 training (both classroom and simulators) in Orlando for 7 days. My first UT System

King Air 350 Captain Khalar

flight as the pilot was on 16 January 2001. I flew System staff to Alpine, Texas.

I learned long ago to always pay attention and to listen carefully to the new people that join your organization. New people see the anomalies and weird things in your operation. I received a package in the mail from the UT System. Opening it, I find my UT System name tag that is supposed to be worn on the chest. I couldn't believe what I see. My social security numbers are on it in large font under my name! No way, and I drop it into the paper shredder.

I week later, I attended the new comers Human Resources briefing. There were about 40 new employees attending. The HR representative covered all the usual issues, then said, "By now you have received your system name tags." I raised my hand and said, "I looked at it, saw my social security number on it, and dropped it into the shredder. I'm not wearing my social security number on my chest." He was taken back and responded, "But you have to turn it in when you leave." The meeting over, the head of personnel came to me and said, "Your right. I'll get you a badge without your numbers. We just started working to eliminate social security numbers from badges."

I was the only employed pilot. While it is legal to fly the KA200 single pilot, I always flew with a second pilot. It was safer, and safety was the name of this game. I hired contract pilots, paying them a day rate. I was extremely particular with who I allowed in the cockpit. Second pilots were paid $315 a day. When I was unavailable to fly, we hired a left seat pilot at $550 a day. Plus, we paid mileage to/from their house, meals, and for over nights, the hotel.

I kept a list of 12 local pilots, male & female. They were all in various endeavors. Some were fractional jet pilots, some corporate pilots, and some owned businesses. I once flew with a Southwest airline pilot. This group was great. One of the female pilots had flown a single engine aircraft around the world in about 7 days. Most were instructor pilots. I even flew with a pilot that I had gone through Air Force pilot training with back in 1969.

When I received a flight request, I would start to contact copilots on my list. I usually called those who made their livelihood by flying strictly day-rate, they needed the work. Then I continued down the list until someone would accept the flight.

Life always presented me with special opportunities. Dallan had been an AFROTC cadet of mine at the University of Texas at Austin while I was the ROTC commander there. He graduated, was commissioned, and eventually became a U2 pilot. Dallan grew up across the street from me. He was home on vacation. A UT System flight to Port Aransas scheduled. I asked Dallan to fly. He agreed and flew with me as copilot. Years later, I flew his mother on a Cirrus22 flight. Dallan is now an airline pilot!

One flight to Houston Hobby had only one passenger. He was the Chancellor, the man who hired me. He returned to the plane for the flight back to Austin. It was day, the weather was perfect, and it would be a short flight. I offered him the copilot seat for the flight home. Wearing a big smile, he hopped into the seat. The hired copilot helped get him strapped in then sat in back.

In between radio calls, I explained to the Chancellor what I was doing, why, and also, who I was communicating with. When we were on final approach to land at Austin, I asked him if he noted that when I pushed forward on the controls, the trees got bigger, and when I pulled back, the trees got smaller. He remembered the hiring interview and laughed. This man hired me, I wanted to thank him. I didn't mention to anyone that I had him in the cockpit with me because I knew my food chain wouldn't be happy.

There are 3 types of flying rules. We flew under FAA Part 91 rules, there were Charter aircraft rules, and then the Airline operation rules. There were significant operational differences. I took off from Austin on a flight to Houston Hobby. We picked up passengers in Houston then flew toward Midland. When we were changed to the Midland approach control frequency, we heard an airliner being given holding instructions. The copilot and I looked at each other. The preflight weather check said it would be clear. We tuned in the Midland weather frequency.

We hear current weather was fog with zero-zero visibility. Time to plan. I asked the copilot to check the weather at airports behind us and at airports past Midland. Looking at the fuel gauge, I said, "We have enough fuel for two approaches before we need to divert." There was a light twin engine aircraft in front of us. It flew a instrument approach and made a missed approach. We descended on the instrument approach to 200 feet above the ground, I saw nothing, and made a missed approach. We followed the light twin for a second approach. This time the twin landed. At 200 feet decision height I saw the runway and we landed.

We operated and flew under FAA, Part 91 rules, so could fly an approach to minimums with zero visibility. The charter aircraft at Austin couldn't depart Austin with the Midland reporting zero visibility. The airliner couldn't make an approach. It had to return to Dallas for fuel upload.

When I hired on, all coordination and flight scheduling was done on paper. The coordination paperwork would be literally walked back and forth across the street between System buildings. Sometimes I would get phone calls late in the evening about the next day's flight. I suggested that an online computer-based scheduling system would improve the operation. A system where everyone could see the multi-month schedule, make flight requests, see the approval process, and would generate a flight manifest. A meeting was held with the System computer folks. I explained what we wanted, then everyone attending the coordination meeting explained what they needed to input and the products they needed to receive. The programmers went away.

A few weeks later they completed an online computer scheduling website. We used the program for a few weeks. Then we had another meeting with the programmers to tweak the program. The final product was awesome! I could log in and see the entire schedule. I could download and print the passenger manifest. Post flight I updated the data with what actually occurred. It could generate Texas State required reports. To my knowledge it is still being used.

We were a pretty smooth-running operation. The secret was Amy. She coordinated the flight schedule and worked any issue that might surface. She was my tether, we communicated day or night via phone, email, and computer. Since I was always on a flight, I only made it to the down town office once every week or two to turn in paperwork, and fuel and expense receipts. Amy's efforts made the operation look good, she is my hero.

President Bush senior agreed to speak at a University of Texas Saturday graduation ceremony. The plan was for us to fly over to Houston Hobby, pick him up and fly him back to Austin, and then return him to Hobby.

Having been in the military for 30 years I knew a thing or two about security. His name appeared on the schedule which was seen by many folks. I asked the schedulers to remove his name from the schedule. When I was asked why, I said I didn't want to become a target.

On the fly day, the senior UT Austin legal representative rode with us. We picked President Bush up along with a Secret Service agent and flew them to the Austin airport. They unloaded. I was in the pilot seat finishing the checklist. I can see cars and additional security men on the ramp out the window.

President Bush

The university legal lady sticks her head in the cockpit and said, “Come on, we are taking pictures.” I followed her through the plane. As I exit, I see my copilot straightening President Bush’s tie. We stood next to the plane taking pictures. Then he left.

A couple hours later he returned and we flew him back to Houston. After parking, the copilot and I sat in the cockpit. President Bush’s head appeared in the cockpit and he said, “Thank you, and I’m sorry for screwing up your weekend.” He didn’t need to talk to us. He came across as a warm, thoughtful man.

While most of the flights were out and back in Texas, I have also flown trips to San Francisco, Kansas City, New Orleans, and Pittsburg. While most of my passengers were System personnel, I also flew the University of Texas President and staff.

Occasionally I flew the President of the University of Texas to away football games. When a trip was scheduled, I'd call the President's office and ask if the pilots would be getting game tickets. I would say I'm not asking for them, but I needed to know to plan ahead. If the answer was yes, and it usually was, then I asked for the crew to ride with the passengers to/from the game. I always wanted to leave and arrive at the plane before or with the passengers. Ground travel was usually with a small bus or limousine.

A couple of trips come to mind. Both were flights to College Station to attend longtime rivals Texas versus Texas A&M football games.

I was scheduled to fly the UT President, his wife, plus staff members to College Station. We flew early in the morning so they could attend an early event. We arrived at the Easterwood Airport to find dense fog (zero/zero visibility) covering the area. We flew the instrument approach down to 200 feet above the ground. Not seeing the runway, I made a missed approach and climbed to 4,000 feet. Approach control directed us into a holding pattern. We were the first plane to arrive that morning. Soon another 10 aircraft arrived, tried to land, but also ended up in holding.

When you are the pilot you have to stay ahead of the aircraft. I had a plan. My personal landing fuel minimums was more than required by the FAA. I always planned to land with one hour's worth of fuel onboard. We checked the weather at the surrounding airports, Waco looked good. I planned to hold until I needed to depart for Waco to upload fuel. I would get there with my 1-hour fuel reserve.

After holding about 40 minutes a passenger head appeared in the cockpit. She was the senior lawyer for the university and asked how long we could hold. I carefully explained the plan and told her that we had one hours holding fuel left, then we would fly over to Waco, land, refuel, and then return. Sometimes in life, the message gets confused. She went back and told the passengers that we only had one more hour of fuel remaining. As in, we would run out of gas in one hour!

The President's wife was a white knuckled passenger. As the holding time went by she apparently became concerned. Thirty minutes went by, she believed we would run out of fuel in the next 30 minutes. Visibility began to improve and we flew an instrument approach and landed.

After landing, the President told me she was counting the time down until we were out of fuel. He kept telling her that I knew what I was doing. I explained to them what the plan had been and that we were safe through the entire flight. I wanted to make sure they knew they were safe.

Another flight to College Station occurred on 24 November, 2011. This was a special game. It was the last football game between Texas and Texas A&M because A&M was leaving the Big 12 football conference.

We rode with the passengers in a limousine to Kyle Field and dropped them off. They had a pregame event to attend. The Limo driver then took the two of us, the copilot and myself, to a restaurant for dinner, then returned us to the stadium. We had tickets with seats in the end zone up in nose bleed country. Our passenger's seats were down on the 50-yard line. This was the usual seating arrangement.

I had a rule. I always wanted to be at the vehicle with or before the passengers, so we always left the game at least 15 minutes before the end. I violated my rule on this trip. This was the last game between these schools. The score was tied 25 to 25 in the 4th quarter during the final seconds. I decided to watch the game to the end. In the last second, the Texas kicker kicked the football through the uprights making the score 25 to 27, and Texas won! I took a picture of the kick.

Game over, the two of us now walked with the shoulder to shoulder crowd down the winding ramps to ground level, then walked around the stadium to where the limo was parked. But no limo! I called the driver. He said they were at the airport terminal. He couldn't come back for us because all the traffic was moving away from the stadium.

Break a rule and pay the penalty. The two of us made the 2-mile walk (half run) back to the airplane and waiting passengers.

United Nations, revisited

I'm going to go back in time and tell you the story of how I learned more about my own history. You read earlier about my 1979 assignment to the United Nations Truce Supervision Origination as an Unarmed United Nations

Military Observer. I left the UN June 1980. From there, I went on to serve in 8 various assignments and locations throughout in the Pacific and in the US.

In 1994, I was on the faculty of the Air War College, Maxwell AFB, Montgomery, AL. I was a seminar leader. One day, I was sitting at my office desk working on a seminar lesson plan. A front office guy appeared at my door and tossed a small plastic package to me. He said it just came in the mail for me and he leaves. I open it, find the coin, and see the words, "1979 United Nations Peace Medal." While it appeared to be something to do with my UN days, there was no letter of explanation as to what it was nor where it came from. Pressed for time, I threw it into a desk drawer and went back to class prep. I didn't think about the coin for years.

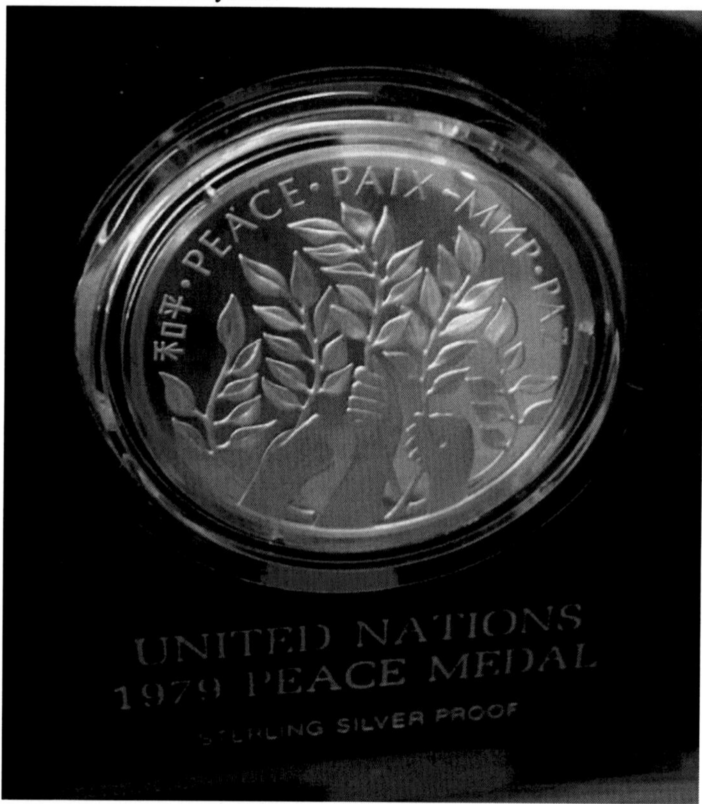

1988 Nobel Peace Prize

Life goes on and after 30 years in the military I retired in September 1999.

A few years later, 2001, I happened to read somewhere that the UN Peacekeepers and Military

Observers had been awarded the Nobel Peace Prize in 1988, thirteen years earlier. Competition was pretty stiff that year. There were 97 candidates including President Reagan, Mikhail Gorbachev, and Pope Paul II.

I did a few google searches and researched it. Then I read the words, "United Nations Peace Medal." The light bulb went on and I realized what I had received years earlier in 1994. I dug around my old military boxes and found my coin, my part of the 1988 Nobel Peace Prize.

In 1988, the Nobel Peace Prize was awarded to all military personnel who served in the UN from 1948-1988. I received the Nobel Peace Prize! Ok, so there were about 500,000 other UN personnel eligible. So that means I own about 1/500,000th of the Nobel Peace Prize. I'll take it. We each received the silver coin, the United Nations Peace Medal.

Many unarmed UNTSO observers and UN armed peacekeepers have died through the years. The last US officer killed was Marine Lt Col Higgins. He was captured near Tyre, Lebanon, tortured, and hung. It's ironic that he died the year (1988) the UN was selected to receive the award. I think about him every time I look at the medal.

Other Opportunities

When I started at UT System, I usually flew twice a week so had a lot of time on my hands. Later the flight frequency increased. The man that told me about the UT System pilot opening approached me with an offer. He was putting together a team that would create a fractional jet company. He wanted me to be the Vice President of Operations. We assembled a 5-person group involved in creating the company named "Time and Travel Solutions, Inc." We had frequent meetings that focused on developing the business model.

The fractional aircraft concept wasn't new. A number already existed such as NetJets & Flight Options. These companies operated large fleets of small business jets. Customers purchase a fraction of a jet. That entitled them to schedule trips in any of the fractional company jets. This was a rich man's world. Our idea was to build up a fractional company, then after 5 or 6 years, sellout to one of the other big operations.

Our plan was built around using two Israeli built aircraft, the Galaxy and the Astra SPX. We drove up to meet with Galaxy Aerospace folks at Alliance Airport in Fort Worth. I flew both planes. Eventually, we arranged a multi-aircraft purchase agreement.

Needing to find a couple million dollars in startup money, we briefed a number of possible investors without success. Finally, a banking institution said they would loan us the money if we had contracts to buy aircraft. We now had aircraft contracts.

I was working the fractional project while I was flying full time for the System. The fractional project was looking good, we had a business concept, aircraft contracts, and bank financing, or so we thought. I was within two weeks of resigning from the UT System and go totally fractional when the bank refused to pass the money to us. At that moment the entire concept fell apart. This was one of my decisions that didn't work, but ultimately, was for the better.

My idea friend rewrote the fractional concept significantly by changing the aircraft to the smaller aircraft, the Cirrus. This company was called "PlaneSmart." I would be the Chief Pilot. One significant difference between business models was that the jet fractional owners would only be passengers who were flown by fractional company pilots. In the Cirrus model, most fraction owners would be pilots and would fly the planes themselves.

The Cirrus was a relatively new aircraft. It was a single engine, propeller driven, 4 seat aircraft. This state-of-art plane had a novel feature. The plane had a parachute that

Cirrus Instructor Pilot

the pilot could deploy for emergency letdowns. It also had outstanding flat panel avionics.

I needed to quickly become an expert in the Cirrus. I studied the flight manuals. The first Cirrus was delivered to the PlaneSmart office located at the Austin airport. I flew it for a couple hours on 23 September 2004. On the 29th, I flew it north to the Cirrus Aircraft Company at the Duluth, Minnesota, airport. There I completed an official aircraft checkout, then completed the Cirrus Standardized Instructor Program. I was now recognized by the Cirrus Company as an expert Cirrus Instructor Pilot.

Returning to Austin, I trained the new PlaneSmart fractional owners in the Cirrus. I was paid $1,800 to give them 10 hours of transition inflight training. I could usually complete a competent pilot in two days, a few took three days, one was so scary and I refused to complete him. I trained about 40 pilots. I recall when I was a young man when my best day working in the woods peeling pulp was $21.

Over time, PlaneSmart relocated the operation from Austin to Dallas. I left the company a short time later. I continued to fly as an independent Cirrus Instructor Pilot in the Austin area for awhile. I charged $500 a day instructor fee.

Plane Change

The System King Air 200 was really old and needed to be replaced. Our guidance from the System Chancellor was no jet and no new aircraft. A couple years earlier I suggested a King Air 350. My suggestion was rejected. They then hired a consultant who, after careful analysis, suggested a King Air 350. The search for one began.

A King Air 350 is a larger aircraft than the King Air 200 so I needed to get special training and pass a check know as a FAA "Type Rating." There were three flight training companies that trained to a KA350 Type Rating. SIMCOM, was in Orlando. I had accomplished my initial King Air 200 training there, then completed annual refresher training there. To complete a KA350 type rating at SIMCOM, I would need to provide an aircraft to complete the check flight. We didn't have one yet. The Flight Safety program was three weeks long. The third option was SimuFlight in Dallas. They said that since I had so many

King Air 200 hours, they could complete me in 6 days! We selected SimuFlight.

I trained at SimuFlight 23-28 July 2008. I realized quickly that it was a big miserable mistake. The course was too short, and way, way, too compressed. It was intense, I spent every minute of my time reading and preparing for the next event. There were classes, then simulators. Finally, it was check ride time.

I failed the check ride! It was the only check ride in my life that I have ever failed. There are two throttles. On the left side of the left throttle is a go-around button. It could be pressed with the pilots' thumb. On an approach, at minimum altitude, if the runway environment isn't seen, you press the go-around button as a missed approach is initiated. The Attitude Indicator then changed to command a climb. I wrapped my hand toward the right side to avoid the left button during the takeoff. What I didn't know was that there was also a go-around button on the right side of the right throttle. Just after takeoff, I inadvertently activated the right-side go-around button. The Attitude Indicator changed to command a climb.

The evaluator froze the simulator, leaned forward, and advised me, "You failed the takeoff task, do you want to continue the check ride or stop?" I said, "Continue." I then completed all the remaining tasks satisfactorily. The fix was one short retraining simulator and then another takeoff check which I passed. (BTW, the plane we bought only had a go-around button on the left side.)

King Air 350, Can Carry 9 Passengers

As I drove south to Austin with the required King Air 350 "Type Rating" in hand, I mentally deleted the training event. The SimuFlight simulator had Pro Line 4 avionics, the King Air 350 that the System bought had newer Pro Line 21 avionics. The cockpit instruments were totally different. Back in Austin, I asked to fly the new plane at least twice before carrying passengers. I needed to operate the new avionics before flying passengers.

I've had people tell me that the Pro Line 21 avionics system was a too complicated to learn without instruction. I did! I learned the Pro Line 21 avionics system by myself. The King Air 350 was awesome and a real pleasure to fly.

Sunset view from the office

On 1 Dec 2008, the System sold the old King Air 200.

When you fly, the situation can change in the next second. I dealt with the usual flying environment issues: maintenance problems, schedule changes, and bad weather. I only experienced one serious malfunction in 14 ½ years.

One night, I flew to El Paso to pick up a University of Texas football coach and fly him to the Redbird airport in Dallas. It was an awesome night flight with a clear star-studded sky. We descended to 4,000 feet just to the west of Fort Worth/Dallas. It was a smooth beautiful ride flying over the Dallas city lights to the Redbird airport. The plan was to drop off the coach, then immediately fly back to Austin.

After I landed, the awesome flight turned to hell when I applied the brakes. My brake pedals went to the floor! I said to the copilot, "I have no brakes. Do you have any brakes on your side?" "No, he responds." I go into full propeller reverse.

We stopped about 500 feet from the runway end. I had him tell the control tower that we were broke down on the runway, to close it, and that we needed to be towed to parking.

We climbed out of the plane to see brake fluid covering the left landing gear and under the wing. The plane was towed into parking. Our drop and go turned into an overnight for me. My copilot had another contract flight the next day and needed to get back to Austin. It turns out that there was only one more commercial flight back that night. I hustled to take him to Dallas Love airport and just made it in time. The next morning, I coordinated a repair for the brakes. It took all day to fix them. I flew it back to Austin late that night.

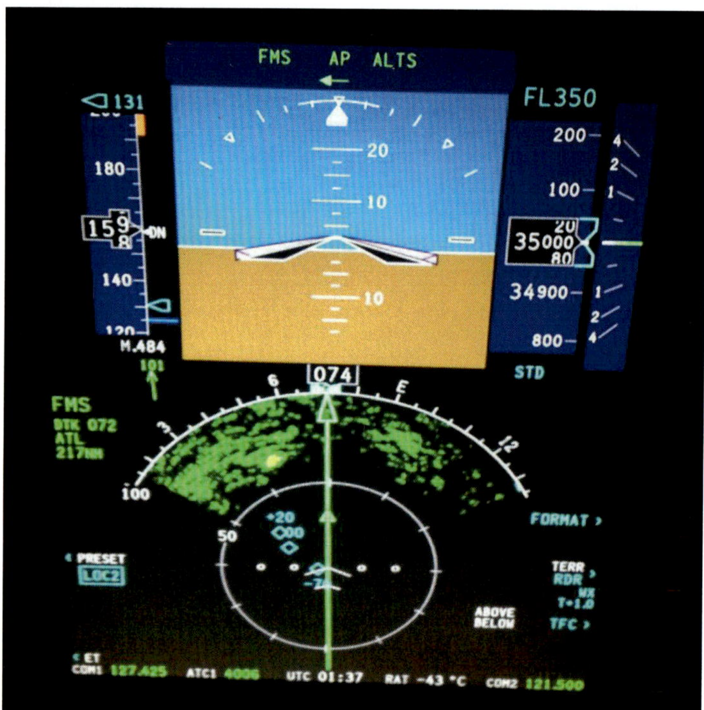

Heading 074, 159 knots, 35,000 feet (Maximum Altitude)

Job Offer

One of my contract pilot buddies asked me for a favor. He flew a King Air 350 for a family out of Georgetown, TX, airport. They were wealthy company owners in the Austin area. He flew them on a day rate basis. Typically, he would fly them out to wherever, and then they would airline him back to Austin. When they wanted to return, they would then airline him back to them for the flight back to Georgetown.

On one flight, he wasn't available to fly them for the return trip, so he asked me if I could I fly them back? The return flight would be from the Marana Regional airport, near Tucson, AZ. I coordinated the request within the University of Texas System and got approval for this one-time flight.

The aircraft owner had been a pilot but had health issues. He was used to sitting up front in the copilot seat and flying his airplane. I agreed to cover the flight for $500 with the stipulation that I would be the only one flying the plane.

The plan was for me to airline out early in the morning on 25 May, 2011. From Austin I would change planes in Dallas, land in Tucson, then take a cab to Marana Airport. I would then fly them, taking off about 11 am in their King Air 350 from Marana back to Georgetown.

I was sleeping when the phone rang about 1145 pm. It was a robo call telling me that my early morning flight was cancelled and rescheduled the next day. What? I got up and quickly checked the reservation on the computer. Yep, it was rescheduled 24 hours later! What to do? I could have called them in the morning and told them about the delay.

I had given my word to fly them the next day. In my world, when pilots agree to fly, the promise was carried out. For 14 ½ years I hired pilots to help crew my flights. I never had a no show. I needed to at least try, so I decided to go to the airport really early to see what was happening and see if I had any options.

It turns out that last night there were severe storms in the Dallas area. Many Dallas flights were cancelled and arriving flights had diverted, many into Austin.

When I arrive at the Austin airport, there were hundreds of people in lines waiting to get to a ticketing counter. The lobby was totally packed. I decided to bypass these long lines and went immediately through security. I got into a short 8-person line at one of the gate counters.

Standing in line, I call the man I'm to fly back to Georgetown. He was someone I've never met or talked to before and told him the situation. I informed him, at that point, I'm not sure when I will get there.

I got to the counter and was scheduled for two evening flights that would get into Tucson about 11 pm. I called my passengers and updated them. If this timing actually occurred, we wouldn't get back to Georgetown until 2 or 3 am. As I walked through the terminal, I saw a flight loading to fly up to Dallas. I walk up to the loading agent and asked if I could get on it. He looked at my tickets, said yes, and I got onto the plane.

So, now I will get to Dallas by 8:30 am, but still had a second flight from Dallas to Tucson arriving at 11 pm that night. Now I expected to spend the entire day in the Dallas airport and leave on the late pm flight.

I landed in Dallas Love airport. As I exited the gateway, I stopped at the gate counter. I explained to the agent that I was trying to get to Tucson but the system had screwed me over. Without any more discussion she took my late evening boarding pass and started typing into the computer terminal. She handed me a new boarding pass saying, "Your flight to Tucson leaves in 40 minutes." I gave her a big," Thank you!"

I walked down about two gates to find my rescheduled flight. The plane was full and there were 72 people on the standby list! Many of these were folks were stranded over night from last night storms. The lady really did me a favor. If she had not printed me a boarding pass, I would have been at the end of the 72-person standby list. I felt a bit sheepish, and certainly didn't mention my new boarding pass to anyone.

I called my passengers with my new arrival time. I arrived in Tucson, then took a $70 cab ride to Marana airport. After I walked through the airport terminal, my passengers were standing next to their King Air 350. I called out, "Do you folks want to fly to Texas?"

The wife went out to pay the $70 taxi fee while I walked around the plane to preflight it. We finally took off. During the flight the owner and I had an interesting talk. He wanted to hear about my background. Then I asked him about his.

One story comes to mind. He had been in the army for a few years. He caught a hop on an Air Force C119 cargo plane. The engines began to malfunction and the crew didn't think they would be able to keep it flying. The loadmaster strapped him into a parachute and threw him out of the back end of the plane. It turns out the crew got the plane to a runway. They had to send out a search party to find him.

After awhile this man who has known me for less than 3 hours totally surprised me with a full-time pilot position offer! I was not expecting that. That meant he wanted to hire me over my pilot friend. The pilot they had was a dedicated, superb pilot. I wouldn't fly him with me in the System plane if he wasn't. They paid him by the day. The offer included medical coverage, and cars. I thanked him for his generous offer but said, no, and that he already had a great pilot.

My job with the UT System was great. The System was a bureaucracy with structure, processes, and policies. If a pilot flies for a family, they owned him. I have heard terrible stories.

We landed at the Georgetown airport. After I shutdown, the wife asked, "Where did you park your car?" I said, "It's down at the Austin airport." They drove me down to the Austin airport parking lot where they dropped me off. Both got out of the car and thanked me for my "never give up" effort in getting to them.

These folks deal with cash. She gave me the agreed $500, then an extra $100. He waved and said he would be happy to fly with me anytime. I waved back.

I never told my friend that I had been offered a fulltime position.

University of Texas System Retirement

I flew for the UT System for 14 ½ wonderful years. I considered my passengers as family. Safety was my primary concern. I always knew when it was time to move on. I decided to retire on my wife's birthday, 30 April 2015. I turned 68 years old on 10 April. I gave the UT System a 9 month heads up to hire replacements. The man I replaced was 70 when he retired so there wasn't a push for me to retire.

Years earlier when I applied for the King Air 200 pilot position with the paper application there were 11 total applicants. Times changed. We advertised on various social media sites and 80 applied! Human Resources asked us to stop taking applications. There were five members on UT System hire team. I was the only pilot. The team was tasked to select 10 from the 80 to interview. I went home with my copies of the 80 applications! What to do? I made three piles of applications on our dining room table. One pile was applications that made no sense or they didn't have a King Air 350 Type Rating. The second pile was of King Air 350 Typed pilots but weren't currently flying a KA350. The last pile was with King Air 350 Typed pilots who were currently flying them. I selected 10 from the last pile.

The hire team met, compared notes, and amazingly, almost all had selected the same 10! We each submitted questions that we thought needed to be asked each candidate during an interview. A question list was created including which of us would ask each question.

We interviewed eight candidates in person, and two via skype (California and Afghanistan). The System decided to hire two pilots. The two we selected had Army flight experience. After the hiring, my running joke (time to screw with them) was that, "It took two Army pilots to replace one Air Force pilot."

The transition was quick and smooth. I had a few months until retirement and flew with them to familiarize them with the System operations. They are both outstanding pilots. I was comfortable knowing that I left my System family in safe hands.

I have from time to time mentioned that I'm a lucky man. I had been hired 14 ½ years earlier. When I applied for the King Air 200 position, I was the only applicant of the 11 with no KA200 experience. According to my three KA350 application piles, I wouldn't have been considered since I had no King Air 200 experience.

My last System flight was on 15 April, 2015. The retirement farewell was on 29 April, and my last day the 30th. (I was the System Chief Pilot for 14 years, 4 months, and 24 days.)

A Farewell gift from the UT System Police Chief

There was a big turnout for my retirement farewell. I had a lot of folks to thank for their continued support through the years. I always considered my System passengers as family. My University of Texas System family gave me a special 50-page book titled, "I Fly Family." My boss, Nancy coordinated it. Members of my family wrote pages of remembrances of our flights together. This book remains very special to me, thank you Nancy.

The man who hired me 14 ½ years earlier, the man I flew illegally in the cockpit with me from Houston to Austin, attended my retirement farewell. I thanked him again for his trust. A few months later I attended his farewell funeral. He died from cancer.

Epilogue

I'm a 71-year-old retired man reflecting back through my life. At the beginning, I mentioned that I simply wanted to document my military experiences for my family and friends. I've done that and more. I wish my parents, and my many uncles and aunts could read this book. Unfortunately, only one uncle survives. I still wish my WWII uncles and military family members had recorded their experiences.

I started this book as a result of a request from a college reunion planning committee. I was asked provide two pages of my Air Force experiences. I expanded those two pages of experiences into this book. While writing, I recalled many more of my life experiences that I didn't include.

I'm comfortable with and proud of my life. I believe I was the first Khalar to graduate from college. We all make decisions on a daily basis that result in significant impact to our lives. Many of my decisions worked, but many didn't. Many of those decisions that didn't work ultimately were positive.

As I think back through my 30 years of experiences in the Air Force, I wonder how I survived. Maybe that is survivor's remorse, "Why them and not me." I mean, we all flew the same missions but some didn't survive. I did. "Band of Brothers" could describe the military experience. It's a world that non-military folks would never understand.

After pilot training, I was assigned as a B52 copilot. Through the years I upgraded to aircraft commander, instructor pilot, and check pilot. Eventually, I became a command evaluator, then tactics officer. Finally, I was a Deputy Commander for Operations and oversaw B52 bomber (19 planes) and KC135 tanker (16 planes) squadrons and 8 divisions totally 380 personnel. I flew over 4,000 hours in the B52. I entered the Air Force as a 2nd Lieutenant and retired as a Colonel.

I fought in 4 wars. During the cold war, I sat on week long alerts with nuclear loaded bombs that were targeted against the Soviet Union. We needed to be able to take off within 15 minutes of notification. I fought in the South East Asia (Vietnam) war. I flew 78 combat sorties dropping just under 5,000 iron bombs on targets in Laos, South Vietnam, and Cambodia. I studied targets in North

Korea and was ready to strike them. During Desert Storm, I was a Chief of Battle Staff in the Strategic Air Command underground command post. We worked B52 bomber and KC135 tanker deployments and operational issues.

I've been around and travelled through 32 counties, and bombed three. (Appendix 6)

After I retired from 30 years in the Air Force, I was hired by the University of Texas System as Chief Pilot. I was there 14 ½ years. The UT System is staffed by unbelievable intelligent and talented personnel. I treated them, my passengers, as my family. I safely flew over 14,000 University of Texas System passengers across 700,000 miles. My last flight was on 15 April, 2015. I retired when I turned 68 years old. I retired with just over 9,000 total flight hours.

As University of Texas System Chief Pilot, I traveled to Orlando, FL, annually to accomplish a King Air 350 refresher course. For three days, in a full motion simulator, I flew flight profiles, while experiencing everything bad (aircraft system failures) that could happen in the plane. I actually practiced to land the plane with both engines out!

On one trip, I stayed in a hotel that was hosting a Nursing Home Operators convention. One evening I had dinner with a few convention members. We had an interesting discussion about aging. We all age. I asked, since they dealt with an aging community, what were their observations of the aging process? How did people change? What sort of life should I expect?

"Well." One member said. And I leaned forward anticipating a profound detailed response about my future aging prospects. He continued, "First, it's go-go, then slow-go, and finally no-go." How profound!

I did the calculations: Renee and I are in the go-go phase! We are always on the go. We are averaging 23,000 miles a year on my car travelling all over the country. Here is a sample of our travels. We have been to the east coast visiting family and friends in Johnson City, TN, and Charleston, SC, and Montgomery, AL. If you ever get a chance to eat Shrimp and Grits, order it. Awesome! New Orleans is always an adventure. Boston is very interesting.

In Boonville, northern New York, we visited Renee's 91-year-old mother, family members, and friends.

On the west coast, we visited my three sons that live in California. Matthew and Pam live in Big Sur State Park where he is a California Park Ranger. I spent a day on patrol with him. I proudly watched him coordinate the rescue of two people stuck half way down a steep cliffside.

Pete, Michelle, and two grandsons, James and Joey, live in Turlock. Pete is an electrician by trade. He plans and constructs large commercial chicken raising facilities across the country. This is a big-time operation. We had a great time with the grandsons.

Chris and Elena live in Santa Clara. Chris works for Comcast keeping folks on line. Elena works at a small zoo. She obviously loves the animals. Elena took me into areas of the zoo where the public doesn't get to go. Fascinating!

My first drive across the country was in September 1970. Since then, I have crossed the country many times. You should travel across this wonderful country. Visit the Grand Canyon, Yellow Stone Park, and the other sites.

We attended the annual Experimental Aircraft Association (EAA AirVenture) airshow in Oshkosh, WI. It's a big event, probably the biggest multi day airshow in the world. I attended the NRA annual convention held in Dallas, TX. It was a great family event.

We winter in our home in Round Rock, TX, and summer at our Highland house in the Town of Highland, WI. My favorite place is sitting on the bank of Sunfish Lake enjoying the wildlife sights and sounds. There are bear, deer, wolves, coyotes, eagles, beaver and the usual squirrels and rabbits.

While in Wisconsin, we traveled back and forth to northern Boonville, NY, by driving across Canada.

My original focus of this book was on my family's military experiences. The Village of Lake Nebagamon built an Armed Services Tribute Wall (Appendix 7) honoring military veterans. Dad and I are on it.

I grow older. I'm currently living the go-go world. I wonder when slow-go or no-go will begin. Then finally, I wonder when the dash will appear as in "Khalar's Dash." (Appendix 8)

Acknowledgements

I began this work admitting I wasn't much of an author. If my high school English teacher (Miss Tartar) was still alive, I'm sure she would agree. I couldn't have completed this without significant help. I have a number of folks to thank for their help with this project.

My wife, Renee, read and made suggestions throughout this work. She also lived some of the events and reminded me about them. Renee helped civilianize it by modifying military verbiage so civilians would understand.

Longtime friends, Janene Jeffery and Marilyn Trainor, were especially helpful. Having no military experience, their editing suggestions were also helpful in my decreasing military jargon making the work more reader friendly. They also helped with grammar and sentence structure.

When I started this project, I made a neighborhood callout for self-publishing suggestions. Richard Bassemir responded. He provided publishing advice. In addition to editing suggestions, he also gave me significant formatting and technical information. If you want to read an interesting story, look for "Chasing Grandpa" on Amazon. Richard wrote about retracing a drive across this country that his grandfather made years ago.

"Band of Brothers" impact each other's lives. Tom Coleman and I served together twice. In the mid 70's Tom was a new B52 copilot who was assigned to my crew. We trained, sat nuclear alert, and deployed together. Years later, in the mid 80's, Tom called and asked me to join him in the 1st Combat Evaluation Group where we evaluated Strategic Air Command squadrons and aircrews. You have read about our time together. Tom read significant portions of this book. His suggestions, comments, and edits were instrumental in helping me to get this story correct. He also reminded me of many events that I had forgotten. Tom, Thanks!

I'm an optimist, my life is good. I want to thank everyone who joined me in my life experiences. My many family members, my many friends, everyone I worked and served with, basically everyone that interacted with me throughout my life. Our interactions made me the person I am. Thank you.

APPENDIX:

1. *Dad's WWII Stories*

Joe Khalar

Dad was a corporal assigned to a Heavy Weapons company during WWII. He was selected to command three half-tracks each armed with four .50 caliber anti-aircraft machine guns. His company commander was a Captain. Dad spoke of him respectfully saying his captain's nickname was "High Pockets." High Pockets and Dad corresponded for years after the war until "High Pockets" died.

While Dad didn't talk much about his WWII experience, he did tell me a few stories.

When I went to war, I rode an aircraft (contract commercial airliner, KC135 tanker, or flew a B52) that would take me into the combat theater in a day. Dad rode in a packed troop ship on a 7 to10 day journey. His ship sailed in a convoy where the speed was dictated by the slowest ship. He made the journey with the threat from German submarines.

He sailed on a Dutch ship. They were fed boiled kidney. Troops were packed in, sleeping in hammocks stacked four high. During heavy seas, many of them became sick. He said vomit sometimes sloshed back and forth across the compartment floor as the ship pitched and rolled. Those were different times.

Dad told me a technique he used while in transit to avoid getting picked for additional duties like KP (Kitchen duty). He said he carried a small clipboard with him. Because of the clipboard, all the senior folks thought he had already been tasked an additional duty so left him alone.

He landed in France at Normandy a few days after D-day and fought across Europe. Dad was attached to General Patton's 3rd Army during the winter drive to relieve Bastogne during the Battle of the Bulge. He told me that every time the armored column stopped, warming fires were started all along the column. It was so cold that sometimes his guns wouldn't fire.

The winter weather was terrible. It was cold and there was a low cloud overcast for over the area for weeks. That meant there wasn't any allied fighter support. Resupply aircraft came in very low dropping supplies. One flew past and other unit gun batteries misidentified it as an enemy plane, fired on it, and shot it down. Dad proudly told me his three half-tracks didn't fire on it.

The armored column that Dad was in slowly moved forward daily. Large formations containing hundreds of B17s flew overhead every day toward targets in Germany. They would disappear for an hour or so, then the survivor formations crossed overhead heading back toward England. There were aircraft strung out due to various performance issues. Many aircraft were damaged, some trailed smoke, and some had one or two engines out. Dad said they laid out large orange cloth panels on the ground so the aircrews could see where the front line was below on the ground. Dad said occasionally crews started to bailout of burning planes once they crossed of the orange panels knowing they were over allied held land. Once a B17 with only two operating engines (one was trailing smoke) flew past at low altitude. There was a German fighter attacking it. Dad said his three half-tracks fired on the German as it passed but missed.

Dad said although they were anti-aircraft guns, they were also effective against personnel. He didn't elaborate as to what that meant but said they fired against ground troops across a river.

He told me an "I shouldn't be alive" story. His armored column was slowly moving along a road in a forest. Looking through the trees, a half-track could be seen about 100 yards away. It was obvious that it had been hit and disabled. Dad stopped his three half-tracks. He wanted to check the destroyed vehicle to see if there was anything on it they could use.

Dad told three crews to cover him while he walked over to the vehicle. The terrain looked flat, but about 50 yards out he came to a deep gulch that ran parallel to the road. Dad looked down into it. About 20 yards away, a Germany soldier was aiming his rifle up at him. They just looked at each other. Dad's arms were hanging at his sides. He made a "move away" motion with his hand and wrist. The German slowly lowered his rifle, looked right and left, then back up at dad, and then he finally trotted off down the gulch and disappeared.

The German could have shot Dad but didn't. Dad could have then shot the German but didn't. He wondered if the German soldier had survived the war.

Dad said that when the war was over, he was in Mons, Belgium, and was assigned to guard a military supply area. He said he happened to be in a local bar and over heard a conversation that his supply area would be robbed the next night. It just so happens that Dad was on duty when the attempt was made. A man was crawling through a cut in the fence. Dad said he put his .45 semi-automatic pistol against the man's head. Robbery was over.

2. *Unofficial Bio*

As a student in college, 1968 AFROTC: I was awarded the Vice Commandants Award as a cadet attending an AFROTC 4-week Field Training at Gunter AF Base, Montgomery, AL. 26 years later I commanded this event.

Graduated 4 June 1969 and went active duty 22 August 1969. Retired 1 September 1999, 30 years! I was a broke college graduate! I departed for Air Force on 22 August 1969. On the way to the Duluth Airport, we stopped at the Poplar State Bank. I took out a 90-day, $120 loan to have money to live on until receiving Air Force pay. Left home entering the Air Force with two suitcases, school loans, and $120 of borrowed money.

Served 30 years and seven days in the Air Force, moved my household 19 times, traveled to 32 countries, and bombed three. Was on duty 24/7, I don't understand the 40-hour work week concept.

During the first week of in-processing at flight school at Randolph, a medical tech took our foot prints. I asked him why the prints? He took us down the hall into the morgue and showed us why. Feet in boots don't burn. The prints were to identify of my body.

I upgraded to B52 aircraft commander with just 1,000 flight hours total flight time. The norm at the time was well over 1,000 hours. My first Aircraft Commander flight with my first crew was a combat sortie over Vietnam.

Flew B52s through the mountains, at night, at 300 feet, training? I guess it wasn't real.

Stood nuclear alert during the cold war living 10 days out of a month in an alert facility at the end of the runway. The bomber was loaded to fight a nuclear war. We were capable of takeoff within 15 minutes of notification, day or night. We studied cold war targets and were capable of hitting our assigned targets.

Flew 78 combat sorties into Cambodia, Laos (What secret war?), and South Vietnam.

18 fellow aviators that were members in my first combat squadron died. There were numerous B52 squadrons across the country with each sustaining similar loss. Were you and yours living a comfortable life? Hope so.

I survived being struck by lightning four times in a B52. Better to be lucky than good!

I landed a B52 that was on fire at Barksdale AFB, Shreveport, La. A fuel line broke. I was the last crewmember out. The firefighters put the fire out on the runway. They are my heroes. Thank you!

I trained to use nuclear bombs and missiles. I dropped 500 lb., 750 lb., and cluster bomb weapons in the real world. I trained to drop naval mines and sink ships with Harpoon anti-ship missiles.

Developed B52 flight tactics and tested equipment. I don't recommend trying to land or air refuel a B52 while wearing night vision goggles. When flying low level multi-axis bombing at night, try not to blowup each other!

I was a Deputy Commander for Operations. I oversaw 19 B52s, 16 KC135 tankers, and 380 crewmembers in the Bomb Wing. I had aircraft and crews were deployed around the world. I didn't lose a single crewmember!

My longest B52 flight was 20 hours from Rome, NY, to Guam. We held for two hours off the Alaskan coast waiting for air fueling from a KC10 tanker. The tanker's takeoff was delayed because of ice fog!

I went to Diego Garcia with a team and accepted the new Air Force facilities from contractors.

In the Middle East, I lived on the West Bank renting from an Arab family. I walked the streets of old Jerusalem.

Chased by the PLO in southern Lebanon while serving as an unarmed United Nations Military Observer in the UN Truce Supervision Organization. They were armed and intended to hurt me. Ironically, I went to another PLO member who protected me. In Lebanon, I saw .50 caliber machine guns, mortars, and artillery fired with intent to kill.

I received the 1988 Nobel Peace Prize! Go ahead and Google it! I served in UNTSO. Ok, there were 500,000 UN members eligible. So, I own 1/500,000 of the 1988 Nobel Peace prize. So, how much are you entitled to?

I rode camels in Egypt, Israel, and elephants in Thailand. The fences in Darwin kept alligators out of the town.

I rode the night train along the Nile River from Cairo to Luxor. I toured Karnack's Temple and King Tuts tomb.

Rode in a cockpit of an Egyptian airliner on a flight from Cairo to Luxor. Think about that! An airliner cockpit!

Saw my youngest son, Chris (at age 6 weeks) for the first time, through a wall of bulletproof glass.

When the North Korean guards were not looking, I stepped across to the North Korea side of the DMZ. I was in North Korea! I have the picture! There are places that the air feels thick with tension, this was one of them.

I met Philippine President Marcos in Hawaii; flew President Bush (senior) in the UT System plane.

Spent a day submerged in the Los Angeles fast attack submarine. I was landed on and catapulted off the aircraft carrier Carl Vincent along the Iranian coast. Spent three days aboard, fantastic!

Have you been to Port Moresby, Papua New Guinea, Canberra or Darwin, Australia, Beirut/southern Lebanon, Cairo/Luxor, Egypt, Larnaka/Pafos, Cyprus, or Riyadh, Saudi Arabia, the Azores, or Jerusalem, Israel? I have.

Stayed at the Khobar Towers, Saudi Arabia. A few years later they were bombed by Muslims.

Was in the infamous "Hanoi Hilton", Hanoi, Vietnam, prison in March, 1999. I was on a military tour! We are friends now with the Vietnamese. I was treated there far better than some of my friends. Torture?

I've been to places that don't exist; done things that didn't happen. Or maybe I'm just too old to remember!

In early 70's I turned down attending Squadron Officer School in residence. This was, at the time, seen as a career killer, oops!

30 years in the military! And I was able to avoid an assignment at the Pentagon. I win! I win!

I don't play golf.

I graduated from the National War College; later served four years on the Air War College faculty.

Served three years as an AFROTC Professor of Aerospace Science at UT Austin, Hook'm Horns!

I commanded a 4-week 520 cadet AFROTC Field Training Camp at Lackland, AFB. On morning, day-1, the Flight Surgeon advised me that a cadet had chicken pox and he wanted to quarantine the entire camp! No way!

BIOGRAPHY

UNITED STATES AIR FORCE

Air War College
325 Chennault Circle
Colonel Richard A. Khalar
Maxwell AFB AL 36112-6427

Colonel Rick Khalar is the Director, Regional Studies Program, Department of International Security Studies, Air War College, Maxwell AFB, Alabama. Colonel Khalar received an Air Force ROTC commission following graduation from the University of Wisconsin, Superior, Wisconsin. He has held a variety of operational and staff positions at the Squadron, Wing, Joint, and MAJCOM levels. Col Khalar is a command pilot with over 4000 hours in B-52 D/F/G/H models and KC-135 aircraft.

EDUCATION

1969 Bachelor of Science degree in Sociology, University of Wisconsin, Superior, Wisconsin.
1976 Master of Arts degree in Sociology, Pepperdine University, Malibu.
1980 Air Command and Staff College, Maxwell Air Force Base, Alabama.
1990 National War College, Ft McNair, Washington, D.C.
1992 Doctor of Public Administration (ABD), University of Alabama

ASSIGNMENTS

1. August 1969 - September 1970, Student, pilot training, Randolph Air Force Base, Texas.
2. October 1970 - July 1976, Co-pilot, Pilot, Instructor and Stan Evaluation pilot, 744th and 34th Bombardment Squadrons, Beale Air Force Base, California.
3. August 1976 - June 1979, Command Post Controller, 92nd Bomb Wing, Fairchild Air Force Base, Washington.
4. June 1979 - June 1980, United Nations Military Observer, United Nations Truce Supervision Organization, Israel/Lebanon.
5. June 1980 - June 1981, Student, Air Command and Staff College, Maxwell Air Force Base, Alabama.
6. August 1981 - July 1982, Battle Staff Operations Officer, CINCPAC Airborne Command Post, Hickam Air Force Base, Hawaii.
7. August 1982 - July 1983, Military Assistant and Aide de Camp, CINCPAC, Camp H. M. Smith, Hawaii.
8. August 1983 - July 1984, Chief, B-52 Standardization and Evaluation Branch; Chief Training Flight, 60th Bombardment Squadron, Andersen Air Force Base, Guam.
9. July 1984 - April 1985, Chief, Operations and Training, 3rd Air Division, Andersen Air Force Base, Guam.
10. May 1985 - February 1988, Assistant Chief Tactics Division; Assistant Chief Bomber Division, 1st Combat Evaluation Group (SAC), Barksdale Air Force Base, Louisiana.
11. March 1988 - July 1989, Deputy Commander for Operations, 416th Bombardment Wing, Griffiss Air Force Base, New York.
12. July 1989 - June 1990, Student, National War College, Ft Lesley J. McNair, Washington, D.C.
13. June 1990 - May 1992, Chief, Mobile C3 Division, Headquarters, Strategic Air Command, Offutt Air Force Base, Nebraska.
14. June 1992 - July 1994, Director, Leadership Management Studies, Department of National

Security Decision Making, Air War College, Maxwell Air Force Base, Alabama.

15. **July 1994 – July 1997, Commander and Professor of Aerospace Studies, AFROTC Detachment 825, The University of Texas at Austin, Austin, Texas.**
16. **July 1997 – July 1998, Professor, Leadership and Ethic Department, Air War College, Maxwell Air Force Base, Alabama.**
17. **July 1998 - August 1999, Director, Regional Studies Program, Department of International Security Studies, Air War College, Maxwell AFB, Alabama.**

FLIGHT INFORMATION

Rating: Command Pilot
Flight hours: More than 4,000
Aircraft flown: B-52D/F/G/H, KC-135A,

MAJOR AWARDS AND DECORATIONS

Defense Meritorious Service Medal
Meritorious Service Medal with 5 oak leaf clusters
Air Medal with 3 oak leaf clusters
Joint Service Commendation Medal
Vietnam Service Medal with 2 Campaign Stars
Republic of Vietnam Gallantry Cross with Palm
United Nations Medal
Republic of Vietnam Campaign Medal

EFFECTIVE DATES OF PROMOTION

Second Lieutenant	**Jun 7, 1969**
First Lieutenant	**Feb 23, 1971**
Captain	**Aug 23, 1972**
Major	**Oct 18, 1979**
Lieutenant Colonel	**Feb 1, 1985**
Colonel	**Sept 1, 1989**

Retired September 1, 1999. Currently living in Austin Texas.

4. *Rules of the Mess*

1. Thou shalt not be late.
2. Thou shalt make every effort to meet all guest.
3. Thou shalt move to the mess when thee hears the chimes and remain standing until seated by the President.
4. Thou shalt not bring drink or lighted smoking materials into the mess.
5. Thou shalt not leave the mess whilst convened. Military protocol overrides all calls of nature.
6. Thou shalt participate in all toasts unless thyself or thy group is being honored with the toast.
7. Thou shalt ensure that thy glass is always charged while toasting.
8. Thou shalt keep toasts and comments within the limits of good taste and mutual respect. Degrading or insulting remarks will be frowned upon by the membership. However, good natured needling is encouraged.
9. Thou shalt not murder the Queen's English.
10. Thou shalt not open the hangar door (talk shop).
11. Thou shalt always use the proper toasting procedure.
12. Thou shalt fall into disrepute with thy peers if the pleats of thy cummerbund are inverted.
13. Thou shalt be painfully regarded if they clip-on bow tie rides at an obvious list. Thou shalt be forgiven, however, if thee also ride at a comparable list.
14. Thou shalt express thy approval by tapping thy spoon on the table. Clapping of hands will not be tolerated.

15. Thou shalt consume thy meal in a manner becoming gentlepersons.
16. Thu shalt not laugh at ridiculously funny comments unless the President first shows approval by laughing.
17. Thou shalt not overindulge thyself in alcoholic beverages.
18. Thou shalt not question the decisions of the president.
19. When the mess opens or adjourns, thou shalt rise and wait for the members at the head table to take their place or depart.
20. Thou shalt not begin eating a course of the meal before members of the head table.
21. Thou shalt not engage in verbal intercourse whilst another member has the floor.
22. Thou shalt not wear an ill-filled or discolored mess jacket.
23. Thou shalt enjoy thyself to thy fullest.

Grog Bowl Procedures

Any member of the mess who has been observed committing a serious breach of protocol, or violating the rules of the mess, shall be punished in order to preserve the decorum of the mess. Punishment will consist of a trip to the grog bowl. The bowl contains a mildly distasteful beverage.

When the President of the mess directs a violator to the grog bowl, the individual proceeds directly to the bowl, squaring all corners in a military fashion. Upon arriving at the grog bowl, the violator will:

1. Salute the grog and say: "_________ (Name) reports"
2. Fill the cup at least one-half full.

3. Do an about face, raise the cup and toast the mess with: "To the Mess."

4. Drink the contents of the cup (without removing it from lips).

5. Hold the cup upside down over his/her head to show that it is empty.

6. Do another about face to replace the cup.

7. Salute the President with: Sir, the rite is complete."

8. Return in a military fashion to his/her seat.

5. *Holbrook, OB-20*

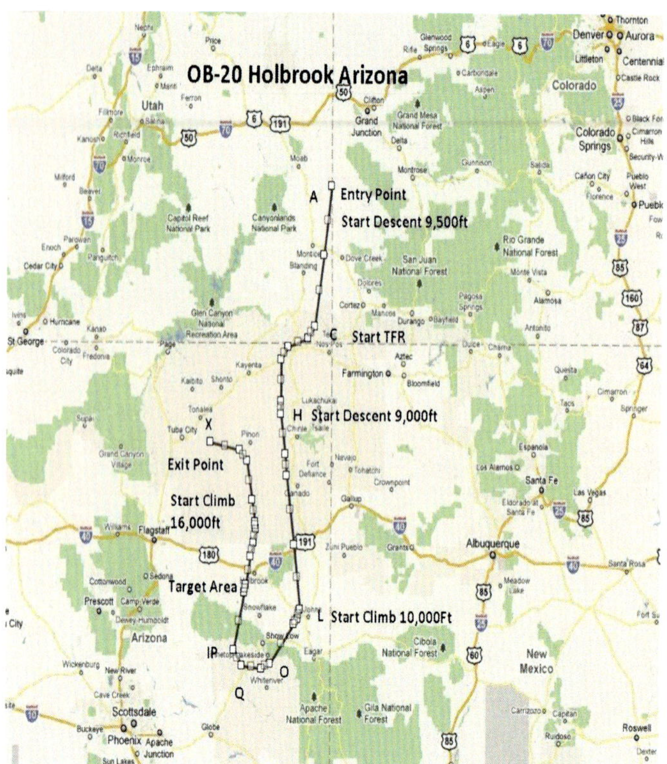

The purpose of the Low-Level Route was to practice flying low enough to be under the enemy radar. Speed 280 knots, accelerating to 325 for the bomb runs. Practice was usually done at 400 feet Above Ground level.

6. *World Travel*

Years	Assigned To	Travel
1969-1970	Randolph AFB, TX	Mexico
1970-1976	Beale AFB, CA	Guam, Thailand, Okinawa (Bombed South Vietnam, Laos, Cambodia)
1976-1979	Fairchild AFB, WA	
1979-1980	United Nations	Israel, Egypt, Lebanon, Cyprus
1980-1981	Maxwell AFB, AL	
1981-1983	Hickam AFB, HI	South Korea, Australia, Japan, Philippines, Papua New Guinea
1983-1985	Anderson AFB, Guam	South Korea, DMZ North Korea, Diego Garcia, Australia, Philippines, Japan, Okinawa
1985-1988	Barksdale AFB, LA	
1988-1989	Griffiss AFB, NY	Guam, Diego Garcia
1989-1990	Ft McNair, Wash DC	Tunisia, Algeria, Morocco, Egypt
1990-1992	Offutt AFB, NE	
1992-1994	Maxwell AFB, AL	Lajes, Saudi Arabia, Israel, Egypt, Jamaica
1994-1997	Un of Texas Austin, TX	Mexico
1997-1999	Maxwell AFB, AL	France, Belgium, Germany, Luxemburg,

		Austria, Vietnam, Singapore, Thailand
1999	Retired	Ireland, Mexico, Canada, Jamaica, Aruba, St Kitt, Puerto Rico

7. *Armed Services Tribute*

On 14 July, 2016, after 5 years of planning, organizers broke ground for the Lake Nebagamon Armed Services Tribute. Construction was completed on the east side of the village auditorium. Committee chairman, Howard Levo, said it was to honor current and past members of the military. The 20-foot wall accommodated 250 veteran tiles. The site includes a handicapped accessible walkway, concrete benches, and recessed lighting that enable visitors to stop by at any hour. I know many names on the Armed Services Tribute wall. I proudly joined my father on the wall.

8. The Dash Poem

I read of a man who stood to speak at the funeral
of a friend. He referred to the dates on the tombstone
from the beginning…to the end.

He noted that first came the date of birth and spoke
of the following date with tears, but he said what mattered
most of all was the dash between those years.

For that dash represents all the time they spent
alive on the earth and now only those who loved them
know what that little line is worth.

For it matters not, how much we own, the cars…
the house…the cash. What matters is how we
lived and loved and how we spent our dash.

So think about this long and hard; are there things
You'd like to change? For you never know how much
time is left that can still be rearranged.

To be less quick to anger and show appreciation
more and love the people in our lives
like we've never loved before.

If we treat each other with respect and more often wear
a smile…remembering that this special dash might
only last a little while.

So when your eulogy is being read, with your life's
actions to rehash…would you be proud of the things
they say about how you lived your dash.